Eco-Innovation in Industry

ENABLING GREEN GROWTH

ORGANISATION FOR ECONOMIC CO-OPERATION AND DEVELOPMENT

The OECD is a unique forum where the governments of 30 democracies work together to address the economic, social and environmental challenges of globalisation. The OECD is also at the forefront of efforts to understand and to help governments respond to new developments and concerns, such as corporate governance, the information economy and the challenges of an ageing population. The Organisation provides a setting where governments can compare policy experiences, seek answers to common problems, identify good practice and work to co-ordinate domestic and international policies.

The OECD member countries are: Australia, Austria, Belgium, Canada, the Czech Republic, Denmark, Finland, France, Germany, Greece, Hungary, Iceland, Ireland, Italy, Japan, Korea, Luxembourg, Mexico, the Netherlands, New Zealand, Norway, Poland, Portugal, the Slovak Republic, Spain, Sweden, Switzerland, Turkey, the United Kingdom and the United States. The Commission of the European Communities takes part in the work of the OECD.

OECD Publishing disseminates widely the results of the Organisation's statistics gathering and research on economic, social and environmental issues, as well as the conventions, guidelines and standards agreed by its members.

ISBN 978-92-64-07721-8 (print)
ISBN 978-92-64-07722-5 (PDF)

Also available in French: *L'éco-innovation dans l'industrie : Favoriser la croissance verte*

Corrigenda to OECD publications may be found on line at: *www.oecd.org/publishing/corrigenda.*

Foreword

The expansion of economic activity in recent decades has been accompanied by growing environmental concerns at the global scale. These include climate change, energy security and increasing resource scarcity. In response, manufacturing industries have recently shown greater interest in sustainable production (sustainable manufacturing) and in undertaking a number of corporate social responsibility (CSR) initiatives. Nevertheless, the incremental progress falls far short of meeting these pressing challenges and improvements in efficiency in some regions have in many cases been offset by increasing volumes of consumption and growth in other regions.

Climate change has become a top priority for OECD governments, and pressure is mounting for world leaders to come up with ambitious medium- to long-term commitments to drastically cut greenhouse gas (GHG) emissions. However, recent OECD analysis suggests that without new policy action, global GHG emissions are likely to increase by 70% by 2050. The political and economic challenges for OECD countries are daunting.

Fortunately, the recent economic crisis has been seen by many as a great opportunity for OECD countries to make the economy stronger and greener. In June 2009, the OECD Council Meeting at Ministerial Level adopted a Declaration on Green Growth. The declaration invited the OECD to develop a Green Growth Strategy to achieve economic recovery in the short term and environmentally and socially sustainable economic growth in the long run.

The OECD is also working towards the completion of the OECD Innovation Strategy, a comprehensive policy strategy to harness innovation for stronger and more sustainable growth and development, and to address the key societal challenges of the 21st century. Innovation will be a key factor in turning the vision of green growth into reality through the development and deployment of environmental technologies and smart solutions. Proactive policy interventions need to steer the course of innovation and encourage industry to take up sustainable practices as business opportunities. Today's policies should aim to stimulate investments not only in promising technologies but also in green infrastructures that facilitate innovative solutions and address long-term societal challenges.

As a contribution to meeting these challenges, the OECD Project on Sustainable Manufacturing and Eco-innovation was launched in 2008 under the auspices of the Committee on Industry, Innovation and Entrepreneurship (CIIE), with the aim to accelerate sustainable production by manufacturing industries as a new opportunity for value creation. This entails spreading existing knowledge and providing industry with a means to benchmark their products and production processes. This project also seeks to promote the concept of *eco-innovation* and to stimulate both technological and systemic solutions to global environmental challenges.

This book presents the research and analysis carried out in the first phase of this project as a part of the OECD Innovation Strategy and the first contribution to the OECD Green Growth Strategy. The following aspects of sustainable manufacturing and eco-innovation were reviewed in order to help policy makers and industry practitioners understand the concepts and practices and to highlight existing gaps in understanding and areas in which further analysis and co-ordination are required:

- review the concepts of sustainable manufacturing and eco-innovation and build up a common framework for analysis (Chapter 1);
- analyse the diverse nature and processes of eco-innovation in manufacturing industries from existing examples (Chapter 2);
- benchmark existing sets of indicators that have been applied by industry for realising sustainable manufacturing (Chapter 3);
- analyse the strengths and weaknesses associated with existing methodologies for measuring eco-innovation at the macro level (Chapter 4);
- take stock of existing national strategies and policy initiatives for promoting eco-innovation in OECD countries (Chapter 5).

Chapter 6 draws together the findings from these research activities and identifies promising work areas for the next phases of the project.

The project has been undertaken by the OECD Directorate for Science, Technology and Industry, and managed by Tomoo Machiba under the supervision of Marcos Bonturi (currently, Office of the Secretary-General) and Dirk Pilat of the Structural Policy Division. The authors of each chapter are:

- Chapter 1: Tomoo Machiba and Karsten Olsen (currently, Amiiko, Denmark).
- Chapter 2: Tomoo Machiba and Karsten Olsen.
- Chapter 3: Kaoru Endo (currently, METI, Japan), Tomoo Machiba and Çağatay Telli (currently, Prime Ministry State Planning Organization, Turkey).
- Chapter 4: Anthony Arundel, René Kemp (both UNU-MERIT, the Netherlands) and Tomoo Machiba.
- Chapter 5: Fabienne Cerri, Laura Chia-Chen Liang, Tomoo Machiba, Lena Shipper (currently, University of Oxford, UK) and Çağatay Telli.
- Chapter 6: Tomoo Machiba.

Hirofumi Oima and Elodie Pierre provided support for the preparation of this publication.

The project has greatly benefited from industry and government insights gained through various opportunities for dialogue, including the International Conference on Sustainable Manufacturing in Rochester, New York (September 2008), two questionnaire surveys and a series of focus group meetings of industry experts. The project's Advisory Expert Group (Chair: Dr. Nabil Nasr, Rochester Institute of Technology) provided useful comments and guidance in the drafting of this volume. The authors would like to thank all participants in those activities and colleagues for excellent support and advice.

This publication is a building block of the OECD Green Growth Strategy. Ministers from 34 countries, including both OECD and non-OECD members, asked us to develop a Green Growth Strategy when they met at the OECD Ministerial Council meeting in June 2009. The aim of the strategy is to provide clear recommendations for how countries can achieve economic growth and development while at the same time moving towards a low-carbon economy, reducing pollution, minimising waste and inefficient use of natural resources and maintaining biodiversity. This entails developing specific tools and policy recommendations across a range of relevant areas from investment and taxes to innovation, trade and employment.

The OECD Green Growth Strategy is prepared through a multi-disciplinary inter-governmental process and is based on the work of the 25 OECD Committees engaged in its development. It will be a fundamental contribution from the OECD to support countries' transition to greener growth in the coming years.

Further information on the Green Growth Strategy is available at: www.oecd.org/greengrowth.

Table of Contents

Acronyms and abbreviations

ADEME	*Agence de l'Environnement et de la Maîtrise de l'Énergie* (Environment and Energy Management Agency, France)
AFV	Alternatively fuelled vehicle
BF	Blast furnace
BOF	Basic oxygen furnace
BSI	British Standards Institution
CAFE	Corporate average fuel economy
CCS	Carbon capture and storage
CFC	Chlorofluorocarbon
CIIE	OECD Committee on Industry, Innovation and Entrepreneurship
CIS	Community Innovation Survey, European Union
CO	Carbon monoxide
CO_2	Carbon dioxide
CRT	Cathode ray tube
CSR	Corporate social responsibility
DCIE	Data centre infrastructure efficiency
DJSI	Dow Jones Sustainability Indexes
DOC	Department of Commerce, United States
DOE	Department of Energy, United States
EAF	Electric arc furnace
EC	European Commission
ECI	Environmental condition indicator
EMAS	Eco-Management and Audit Scheme, European Union
EMS	Environmental management system
EPA	Environmental Protection Agency, United States

EPE	Environmental performance evaluation
EPI	Environmental performance indicator
EPO	European Patent Office
ETAP	Environmental Technologies Action Plan, European Union
ETV	Environmental technology verification
EU	European Union
GDP	Gross domestic product
GHG	Greenhouse gas
GRI	Global Reporting Initiative
GSCM	Green supply chain management
ICT	Information and communication technology
IEA	International Energy Agency
IPPC	Integrated pollution prevention and control
ISO	International Organization for Standardization
IT	Information technology
KPI	Key performance indicator
KTN	Knowledge Transfer Network, United Kingdom
LCA	Life cycle assessment
LCD	Liquid crystal display
METI	Ministry of Economy, Trade and Industry, Japan
MFA	Material flow analysis
MFCA	Material flow cost accounting
MIPS	Material input per service unit
MOE	Ministry of the Environment, Japan
MPI	Management performance indicator
NCPC	National Cleaner Production Centre
NGO	Non-governmental organisation
NOx	Nitrogen oxides
OPI	Operational performance indicator

PACE	Pollution abatement and control expenditures
PATSTAT	EPO/OECD Patent Statistical Database
PC	Personal computer
PRTR	Pollutant Release and Transfer Register
PSS	Product-service system
R&D	Research and development
RTD	Research and technological development
SME	Small and medium-sized enterprise
SO_2	Sulphur dioxide
SOx	Sulphur oxides
SRI	Socially responsible investment
UNCED	United Nations Conference on Environment and Development
UNECE	United Nations Economic Commission for Europe
UNEP	United Nations Environment Programme
UNIDO	United Nations Industrial Development Organization
UNU	United Nations University
UNU-MERIT	Maastricht Economic and Social Research and Training Centre on Innovation and Technology, UNU and University of Maastricht, the Netherlands
USPTO	United States Patent and Trademark Office
WBCSD	World Business Council for Sustainable Development
ZEW	*Zentrum für Europäische Wirtschaftsforschung* (Centre for European Economic Research), Mannheim, Germany

Preface

As the world emerges from the worst financial and economic crisis in recent history, pressure is mounting on world leaders to commit to drastic cuts in greenhouse gas emissions and tackle climate change. Past crisis periods have often served as a springboard for change and the current crisis provides a great opportunity for the global economy to shift track. New policies and frameworks will be needed to restore sustainable economic growth, prevent environmental degradation and enhance quality of life. Innovation will be one of the keys to putting countries on a path to more sustainable, smarter and greener growth.

The OECD is currently finalising an Innovation Strategy for the 21st century, to foster economic growth and to tackle the major global challenges of our time, including climate change. This strategy is adapted to innovation today, which has increasingly becoming global and knowledge-based. Innovators now connect across the planet, through global value chains and networks, enabled by the growing role of the Internet. Governments need to understand these new trends and design their policies accordingly – next-generation innovation policies should take the full cycle of innovation into account and look beyond R&D. Such policies will need to foster the commercialisation of promising technologies and enable non-technological forms of innovation such as service development and organisational changes. When completed in 2010, the OECD Innovation Strategy will help governments devise policies that keep pace with these changes and promote productivity and growth in a sustainable way.

The Innovation Strategy will also feed into the OECD's efforts to support countries in their drive for green growth. In June 2009, the OECD Council Meeting at Ministerial Level adopted a Declaration on Green Growth and endorsed a mandate for the OECD to develop a Green Growth Strategy. In the Declaration, Ministers from 34 countries jointly affirmed that they will "strengthen their effort to pursue green growth strategies as part of their responses to the current crisis and beyond, acknowledging that green and growth can go hand-in-hand."

Innovation will help turn the vision of green growth into reality as it is the key to the development and deployment of environmental technologies and smart solutions. *Eco-Innovation in Industry: Enabling Green Growth* explores the crucial linkages between innovation and green growth. The study reviews current industry and policy practices to foster eco-innovation, and explores existing concepts and measurement methods. More importantly, it examines the policy interventions that will be needed to steer innovation towards sustainable development and encourage industry to take up sustainable practices. It finds that in many leading firms, improvements in sustainability and the bottom line can go together. In the coming years, the OECD will accelerate its efforts to help governments across the globe to identify policies that can achieve stronger, cleaner and fairer growth.

Andrew Wyckoff
Director
Directorate for Science, Technology and Industry
OECD

Executive Summary

The evolution of sustainable manufacturing has been facilitated by multi-level eco-innovation

Manufacturing industries have the potential to become a driving force for realising a sustainable society by introducing efficient production practices and developing products and services that help reduce negative impacts. This will require them to adopt a more holistic business approach that places environmental and social aspects on an equal footing with economic concerns.

Their efforts to improve environmental performance have been shifting from "end-of-pipe" pollution control to a focus on product life cycles and integrated environmental strategies and management systems. Furthermore, efforts are increasingly made to create closed-loop, circular production systems in which discarded products are used as new resources for production.

Many companies and a few governments have started to use the term *eco-innovation* to describe the contributions of business to sustainable development while improving competitiveness. Eco-innovation can be generally defined as innovation that results in a reduction of environmental impact, no matter whether or not that effect is intended. Various eco-innovation activities can be analysed along three dimensions:

- *targets* (the focus areas of eco-innovation: products, processes, marketing methods, organisations and institutions);
- *mechanisms* (the ways in which changes are made in the targets: modification, redesign, alternatives and creation); and
- *impacts* (effects of eco-innovation on the environment).

Innovation plays a key role in moving manufacturing industries towards sustainable production, and the evolution of sustainable manufacturing initiatives has been facilitated by eco-innovation. As those initiatives advance, the process of their implementation becomes increasingly complex and industries need to adopt an approach that can integrate the various elements of eco-innovation to leverage the maximum environmental benefits. Such advanced, multi-level eco-innovation processes are often referred to as *system innovation* – innovation characterised by shifts in how society functions and how its needs are met.

Technological eco-innovations are often complemented by non-technological changes

To better represent the contexts and processes that lead to eco-innovation, some illustrative examples of eco-innovative solutions have been collected from three sectors: automotive and transport, iron and steel, and electronics. The examples were examined in light of the three dimensions of eco-innovation mentioned above.

Many eco-innovation initiatives in the automotive and transport industry have focused on improving the energy efficiency of vehicles while heightening their safety. The iron and steel industry has in recent years introduced a number of energy-saving modifications and has redesigned various production processes. While the electronics industry has mostly been concerned with the energy consumption of products, growing consumption of the products themselves has also led the industry's effort to increasing recycling possibilities. Overall, technological advances tend to be the primary focus of current eco-innovation efforts. These are typically associated with products or processes as eco-innovation targets, and with modification or redesign as the principal mechanisms.

Nevertheless, a number of complementary non-technological changes have functioned as key drivers. Such changes have been either organisational or institutional in nature. They include the establishment of separate environmental divisions to monitor and improve overall environmental performance and help direct R&D efforts, and the establishment of inter-sectoral or multi-stakeholder collaborative research networks. Some industry players have even started exploring more systemic eco-innovation through the introduction of new business models and alternative modes of provision, such as bicycle-sharing schemes and product-service solutions in photocopying and data centre energy management.

The essence of eco-innovation cannot necessarily be adequately represented by a single set of target and mechanism characteristics. Instead, it seems best examined in terms of an array of characteristics ranging from modifications to creations across products, processes, organisations and institutions.

Existing indicators can be applied in combination to accelerate corporate sustainability efforts

Indicators help manufacturing companies define objectives and monitor progress towards sustainable production. Existing indicators for sustainable manufacturing are diverse in nature and have been developed on a voluntary basis or set as an industry standard or by legislation. To analyse their effectiveness for guiding companies' sustainable manufacturing efforts, nine representative sets of indicators were reviewed (individual indicators, key performance indicators, composite indices, material flow analysis, environmental accounting, eco-efficiency indicators, life cycle assessment indicators, sustainability reporting indicators, and socially responsible investment indices) based on six benchmarking criteria (comparability, applicability for small and medium-sized enterprises, usefulness for management, effective improvement in operations, possibility of aggregation, and effectiveness for finding innovative solutions).

The benchmarking results show that there is no ideal single set of indicators which covers all of the aspects companies need to address to improve their production processes and products. Except for eco-efficiency indicators, each of the nine categories is mainly designed to help management decision making or to facilitate improvements in products or processes at the operational level. In reality, many companies are applying more than one set of indicators at different levels, often without relating them.

An appropriate combination of existing indicator sets could help give companies a more comprehensive picture of economic, environmental and social effects across the value chain and the product life cycle. The further development and standardisation of environmental valuation techniques could also help companies make more rational decisions on investments in sustainable manufacturing activities. New system-level indicators may also be needed to identify the wider impacts of introducing new products and production processes beyond a single product life cycle. Small and medium-sized enterprises (SMEs) and suppliers need to start by collecting data for a minimum set of individual indicators and then adopt more advanced indicators step by step.

Different data sources would help identify overall patterns of eco-innovation activities

Quantitative measurement of eco-innovation activities would help policy makers and industries grasp trends. It would also raise awareness of eco-innovation among stakeholders and make improvements achieved through eco-innovation more evident. To explore future opportunities for measurement, the strengths and weaknesses of existing methods of measuring eco-innovation at the macro level (*i.e.* sectoral, local and national) are analysed.

It is important to investigate the nature (how companies innovate), drivers, barriers and impacts of eco-innovation in order to capture the overall picture. These aspects can be captured by four categories of data: input measures (*e.g.* R&D expenditure); intermediate output measures (*e.g.* number of patents); direct output measures (*e.g.* number of new products); and indirect impact measures (*e.g.* changes in resource productivity). Relevant data can be obtained either by using generic data sources or by conducting specially designed surveys.

Each measurement approach has its strengths and weaknesses, and no single method or indicator can fully capture eco-innovation activities. Generic data sources can provide readily available information on certain aspects of the nature of eco-innovation, but it may narrow the scope and aspects of eco-innovation to be analysed. While surveys can enable researchers to obtain more detailed and focused information, they are costly to conduct and the number of respondents is likely to be limited. To identify overall patterns of eco-innovation, it is therefore important to apply different analytical methods, possibly combined, and examine information from various sources with an appropriate understanding of the context of the data considered.

Supply- and demand-side policies should be better aligned to facilitate eco-innovation

Governments in OECD countries have mainly used their environmental policies to promote sustainable manufacturing and eco-innovation, without necessarily building coherence or synergy with other policies. More recently, environmental concerns have started to be integrated in innovation policies. This trend needs to be supported to help achieve ambitious environmental and socio-economic goals simultaneously, as environmental and innovation policies can reinforce each other.

To gain insight into current government policies, existing national strategies and overarching initiatives were analysed based on responses to a questionnaire survey from ten OECD countries (Canada, Denmark, France, Germany, Greece, Japan, Sweden, Turkey, the United Kingdom and the United States). The survey found that an increasing number of countries now perceive environmental challenges not as a barrier to economic growth but as a new opportunity for increasing competitiveness. However, not all countries surveyed seem to have a specific strategy for eco-innovation; when they do, there is often little policy co-ordination among the various departments involved.

Initiatives and programmes that promote eco-innovation are diverse and include both supply-side and demand-side measures. Many supply-side initiatives involve the creation of networks, platforms or partnerships that engage different industry and non-industry stakeholders, in addition to conventional measures for funding research, education and technology demonstration. Demand-side measures such as green public procurement are receiving increasing attention, as governments acknowledge that insufficiently developed markets are often the key constraint for eco-innovation.

Current demand-side measures are often poorly aligned with existing supply-side measures and need a more focused approach to leveraging eco-innovation activities. A more comprehensive understanding of the interaction between supply and demand for eco-innovation will be a prerequisite for creating successful eco-innovation policy mixes.

More OECD work on indicators and case analysis would help advance global efforts

The above outcomes of research and analysis are drawn together into nine key findings (see Chapter 6). Identified together with the project's advisory expert group, promising areas for the work of the OECD project on sustainable manufacturing and eco-innovation in the next phase (2009-10), and possibly beyond, include:

- **Provide guidance on indicators for sustainable manufacturing**: The OECD could bring clarity and consistency to existing indicator sets by developing a common terminology and understanding of the indicators and their use. It could also play a role in providing supportive measures for increasing the use of indicators by supply chain companies and SMEs.

- **Identify promising policies for eco-innovation**: Better evaluation of the implementation of various policy measures would be helpful to identify promising eco-innovation policies. The OECD can also facilitate the sharing of best policy practices among governments.
- **Build a common vision for eco-innovation**: The OECD could help fill the gap in understanding eco-innovations, especially those that are more integrated and systemic and have non-technological characteristics, by co-ordinating in-depth case studies. This could form the basis for developing a common vision of environmentally friendly social systems and roadmaps to achieve this goal.
- **Develop a common definition and a scoreboard**: With the substantial insights obtained, the OECD could consider the development of a common definition of eco-innovation and an "eco-innovation scoreboard" for benchmarking eco-innovation activities and public policies by combining different statistics and data.

Chapter 1

Framing Eco-innovation: The Concept and the Evolution of Sustainable Manufacturing

This chapter presents the notions of sustainable manufacturing and eco-innovation. It explores the relation between them in order to facilitate the analysis of manufacturing initiatives directed towards sustainable development. Every shift in such initiatives – from conventional pollution control and cleaner production to the development of new business models and eco-industrial parks – can be understood as facilitated by eco-innovation. The application of the eco-innovation concept offers a promising way to move industrial production in a more sustainable direction and respond to pressing global challenges such as climate change.

Introduction

The primary goals of a sustainable society concern the creation of material wealth and prosperity, the preservation of nature and the development of beneficial social conditions for all human beings. Interest in creating a sustainable society has been building among politicians, business leaders and the general public. This is particularly evident in the current debate on climate change and the level to which the issue has risen on the global political agenda, especially after the economic crisis which began in 2008.

Manufacturing industries account for a significant part of the world's consumption of resources and generation of waste. Worldwide, the energy consumption of manufacturing industries grew by 61% from 1971 to 2004 and accounts for nearly a third of global energy usage. Manufacturing industries are also responsible for 36% of global carbon dioxide (CO_2) emissions (IEA, 2007). However, these figures do not cover the extraction of raw materials and the use of manufactured products; if they did, the impact would be far greater. To date, manufacturing industries have taken various steps to reduce environmental and social impacts, largely owing to stricter regulations and growing pressure to take more responsibility for the impact of their operations. There is also a growing trend for companies to voluntarily improve their social and environmental performance for reasons relating to higher profitability, increased efficiency and greater competitiveness. As a result, industries are gradually moving from pollution control and treatment measures to more integrated and efficient solutions.

Nonetheless, the urgency of further action to avoid continuing environmental degradation is widely recognised. Improvements in resource and energy efficiency in some regions have often been offset by increasing consumption in others, and efficiency gains in some areas are outpaced by scale effects. The International Energy Agency (IEA) predicts that the global energy-related CO_2 emissions will increase by 25% by 2030 even under the current best policy scenario (IEA, 2007). This emphasises the need to alter patterns of production and consumption so as not to put further pressure on the planet.

Hence, the pressure on manufacturing industries to reduce their environmental and social impacts is bound to increase further. At the same time, they can become a driving force for the creation of a sustainable society by designing and implementing integrated sustainable practices that allow them to eliminate or drastically reduce their environmental and social impacts. They can also develop products that contribute to better environmental performance in other sectors. This calls for a shift in the perception of industrial production from one in which manufacturing is understood as an independent process to one in which it is an integral part of a broader system (Maxwell *et al.*, 2006). This in turns requires the adoption of a more holistic business approach that places environmental and social aspects on an equal footing with economic concerns.

This chapter introduces the concepts of sustainable manufacturing and eco-innovation and considers the possibility of considering the two concepts within a common analytical framework. The OECD hopes that this exercise will facilitate better understanding of current sustainability initiatives in industry and provide guidance on how to encourage future industry activities in this direction.

The following discussion first categorises different notions of sustainable production that have been promoted and applied in manufacturing industries over the last few decades. Second, it gives a conceptual overview of eco-innovation and indicates how this concept may help the manufacturing sector to improve its sustainable production initiatives. Finally, it explores the conceptual relations between sustainable manufacturing and eco-innovation as a means of analysing current initiatives from a broader perspective and spreading good practices in the sectors, especially among supply chain companies and small and medium-sized enterprises (SMEs). The chapter focuses on environmental aspects of sustainable development.

The rise of sustainable manufacturing

The idea of sustainable development emerged in the early 1980s in the wake of growing concerns over the environmental damage associated with economic growth (IUCN, 1980). Today it is typically associated with development that ensures environmental protection, economic wealth and social equity – known as the three pillars of sustainable development – such that the needs of present generations can be met without compromising the ability of future generations to meet theirs (WCED, 1987). The use of "sustainability" in specific areas such as production, manufacturing, innovation, etc., tend to rely on this definition, albeit within a more confined context.

There appears to be no generally accepted definition of sustainable manufacturing but the concept fits well within the broader notion of sustainable production. The concept of sustainable production emerged from the United Nations Conference on Environment and Development (UNCED) held in Rio de Janeiro in 1992 as a vital means of realising sustainable development (Veleva and Ellenbecker, 2001). The Lowell Center for Sustainable Production at the University of Massachusetts, Lowell, defines sustainable production as "the creation of goods and services using processes and systems that are: non-polluting, conserving of energy and natural resources, economically viable, safe and healthful for workers, communities, and consumers, and socially and creatively rewarding for all working people" (Nasr and Thurston, 2006). With specific reference to "production in manufacturing sectors", this provides a good starting point for defining sustainable manufacturing and is used as a baseline here, although, as noted, this chapter mainly deals with the environmental aspects.[1] This section describes sustainable manufacturing initiatives and how these have evolved over time.

The first step: pollution control and treatment

In the past, the environmental harm caused by industrial production was typically dealt with on the basis of "the solution to pollution is dilution", that is, by dispersing pollution in less harmful or less apparent ways (UNEP and UNIDO, 2004). More recently, driven by stricter environmental regulations, industry has mostly dealt with environmental harm by attempting to control and reduce the amount of emissions and effluents discharged into the environment through various treatment measures.

Pollution control is characterised by the application of technological measures that act as non-essential parts of existing manufacturing processes at the final stage of these processes. They are often referred as "end-of-pipe" technologies or solutions (Figure 1.1). In general, the alleviation of environmental harm in this way stems from reducing or removing air, soil, and water contaminants that were already formed in the production process.

Figure 1.1. Pollution control and treatment

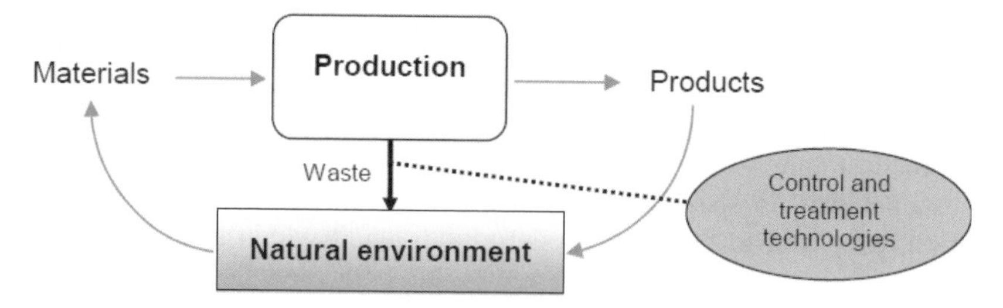

Since pollution control does not restructure the existing production systems in any major way, the only benefit is better environmental performance. Manufacturing companies have traditionally perceived investment in such measures as a costly burden. They typically feel that industrial competitiveness suffers from the costs of environmental protection and clean-up and that environmental performance weighs on profitability and economic growth (Porter and van de Linde, 1995).

When dealing with environmental harm, curative solutions are still essential for most manufacturing industries and their potential impact is far from insignificant. Examples include biological and chemical components for the treatment of waste water, air filtration systems and acoustic enclosures for noise reduction. In the context of climate change, the latest carbon capture and storage (CCS) technologies are also highly relevant.

Working towards preventive solutions and cleaner production

In the effort to shift environmental management from conventional pollution control to a more proactive approach, the United Nations Environment Programme (UNEP) introduced a Cleaner Production Programme in 1989. The concept of cleaner production builds on the precautionary principle, a philosophy of "anticipate and prevent", through an integrated environmental strategy. Since 1994, the UNEP has worked with the United Nations Industrial Development Organization (UNIDO) to set up national cleaner production centres (NCPCs) worldwide to spread the industrial application of this philosophy. By 2007, 37 NCPCs had been established.

The major factor distinguishing cleaner production from pollution control and treatment is the fact that the focus shifts towards earlier stages in the industrial process, *i.e.* the source of pollution. The shift towards cleaner production entails investigating all aspects of the production process and its organisational arrangements to identify areas in which environmental harm can be reduced or eliminated. These areas are often categorised as follows (Ashford, 1994):

- housekeeping, which refers to improvements in work practices and maintenance;
- process optimisation, which leads to the conservation of raw materials and energy;
- raw material substitution, which eliminates toxic materials by shifting to more environmentally sound resources;
- new technologies, which enable reductions in resource consumption, waste generation and emissions of pollutants;
- new product design, which aims to address and minimise environmental impacts.

The concept of cleaner production embraces the notion of efficient resource use while avoiding unnecessary generation of waste (Figure 1.2). Improvements in environmental performance based on lowering pollution at the source require changes to existing manufacturing processes, products/ services, and/or organisational structures and procedures. Even though the implementation of cleaner production stays within the manufacturing company, as is the case with pollution control, it leads to a more integrated environmental approach and is considered essential for moving towards eco-efficient production (see next section). The potential economic and environmental benefits of cleaner production are therefore often superior to those of end-of-pipe solutions.

Figure 1.2. Cleaner production

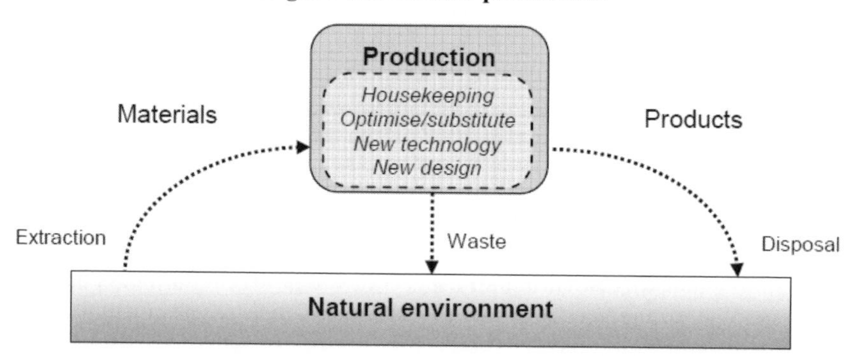

Note: The perspective of the natural environment is broader than for pollution control and treatment (Figure 1.1) as the concept of cleaner production takes the whole production process into account.

The implementation of cleaner production initiatives also constitutes a larger and more challenging task. It may be hampered in particular by barriers within companies that arise from problems of organisational co-ordination as well as insufficient managerial support. Additional obstacles may arise from regulatory environments in which specific technology standards imposed by regulations favour end-of-pipe abatement measures rather than cleaner production (Frondel *et al.*, 2007).

However, a recent survey of more than 4 000 manufacturing facilities in Canada, France, Germany, Hungary, Japan, Norway and the United States (Frondel *et al.*, 2007) shows that more than 75% of respondents reported mainly investments in cleaner production technologies. The data also show that end-of-pipe technologies are typically introduced to comply with regulations, while the implementation of cleaner production technologies is driven by the potential for increasing manufacturing efficiency and reducing costs of operations. This was indicated by a positive correlation between corporate investments in end-of-pipe technologies and respondents' assessment and perception of the stringency of regulatory measures and environmental policies; cost-saving motives and the responding companies' use of specific environmental management tools (*e.g.* environmental policies, accounting, audits, etc.) were correlated with investments in cleaner production.

Managing the transition to eco-efficiency

With the shift from pollution control to pollution prevention, environmental considerations and the improvement of environmental performance in manufacturing industries are also increasingly regarded from the perspective of business interests rather than regulatory compliance. In many cases, companies have found that what is good for the environment is not necessarily bad for business. In fact, it may lead to a competitive edge because of better general management, optimisation of production processes, reductions in resource consumption, and the like (Box 1.1). "Going green" is progressively seen as a potentially profitable direction, and voluntary and pre-emptive sustainability initiatives have become increasingly common in recent years.

Box 1.1. Savings through better environmental performance

The Green Suppliers Network co-ordinated by the US Environmental Protection Agency (EPA) seeks to help SMEs in the manufacturing sectors through programmes that help companies to identify strategies for implementing cleaner production techniques. A review of the results of 60 programmes shows strong evidence of improved environmental performance as well as large savings for the companies. Experiences from European initiatives also show that a considerable number of SMEs are increasingly interested in implementing cleaner production to improve their economic and environmental performance.

Source: Green Suppliers Network, *www.greensuppliers.gov*;
Kurzinger (2004), "Capacity Building for Profitable Environmental Management", *Journal of Cleaner Production*, Vol. 12, No. 3.

A range of developments in the global economy are strengthening the demand for greater efficiency. The globalisation of manufacturing production and its value chain, for example, is strengthening competitive pressures, and the need for manufacturing companies to improve their cost-effectiveness is increasing. Combined with growing resource constraints, which have led to higher costs of core manufacturing activities, incentives to ensure resource efficiency are becoming stronger.

To help companies step up their contribution to the creation of a sustainable society while remaining competitive in the global market, the World Business Council for Sustainable Development (WBCSD) introduced the concept of eco-efficiency, which was put forth as one of industry's key contributions to sustainable development at the time of the UNCED in 1992 (Schmidheiny, 1992).[2]

The WBCSD defines eco-efficiency as a state that can be reached through "the delivery of competitively priced goods and services that satisfy human needs and bring quality of life while progressively reducing environmental impacts of goods and resource intensity throughout the entire life cycle to a level at least in line with the Earth's estimated carrying capacity" (WBCSD, 1996). The goal of eco-efficiency is the adoption of production methods that go hand in hand with an ecologically sustainable society and encompasses a range of other important concepts surrounding sustainable production and manufacturing.

Over the last decade, the original idea and importance of eco-efficiency as a guiding principle for industrial production and business decisions has gained much broader attention and has been promoted with a simple catchphrase "doing more with less", *i.e.* producing more goods and services while using fewer resources and creating less waste and pollution (EC, 2005). This movement has led to a diverse range of conceptual and methodological approaches such as environmental monitoring and auditing and environmental strategies (Maxwell *et al.*, 2006), which companies can use to implement eco-efficiency principles in production.

Such tasks are not trivial for manufacturing companies and place great demands on their organisational management capability. The development of environmental management systems (EMSs) has tied together many of the environmental monitoring and management principles, providing a framework for companies to move towards eco-efficient production (Johnstone *et al.*, 2007).

An EMS is meant to provide companies with a comprehensive and systematic management system for continuous improvement of its environmental performance. Once implemented, the system relies on a structure that is typically characterised by four cyclical, action-oriented steps: *i)* plan; *ii)* implement; *iii)* monitor and check; and *iv)* review and improve (Perotto *et al.*, 2008) (Figure 1.3). These steps are applied across all elements of the company's activities, products and services that interact with the environment (ISO, 2004), and may include the restructuring of processes and responsibilities throughout the company.

Figure 1.3. A typical cycle of environmental management systems

To take account of organisational and industry differences EMSs can be implemented in many ways. Standards nevertheless exist for securing the respect of the main principles. The two main standards, for which a certification also can be obtained, are ISO 14001, developed by the International Organization for Standardization (ISO), and the Eco-Management and Audit Scheme (EMAS), developed by the European Commission. These schemes aim to ensure that companies adopt an environmental policy, that environmental responsibilities are clearly designated throughout the organisation, and that they undergo external audits of the system.

The implementation of an EMS can be useful not only for improving the environmental performance of manufacturing processes (Johnstone *et al.*, 2007) but also for meeting increasing pressures from stakeholders, improving the corporate image, and reducing risks of environmental liabilities and non-compliance (Perotto *et al.*, 2008). Much evidence, albeit mostly from case studies of individual companies, also indicates that the introduction of EMSs leads to better financial performance. The number of EMS certifications has grown substantially in some countries, though the proportion of certified companies is still very low.

The measurement of environmental performance lies at the heart of any EMS as it provides information that is essential for managing and reducing environmental impacts. Assessing environmental performance is not a marginal task, however, and is subject to methodological debates.[3] Environmental performance is typically monitored through process measurements with the help of various indicators that aim to summarise and simplify relevant information from the production system (indicator issues are extensively discussed in Chapter 3).

Life cycle thinking and green supply chain management

Life cycle assessment (LCA) is one of the most widely used tools for measuring environmental impacts and deciding on the development of new products and processes. As the name suggests, its aim is to reduce the use of resources and environmental impacts throughout the entire life span of products and services. Life cycle thinking goes beyond cleaner production as it emphasises the need for companies to look beyond conventional organisational boundaries when considering the environmental impacts of their activities. This involves taking into account environmental impacts and responsibilities that arise from the extraction of materials through the design of products and production processes to the consumption and the final disposal of products. For this reason, LCA is also referred to as "cradle-to-grave" analysis.

The life cycle philosophy and management approaches have laid the foundation for a range of relatively new and proactive environmental initiatives and business models, in which environmental considerations go beyond the manufacturing facility to the entire value chain. On the policy level, this trend is reflected in Extended Producer Responsibility initiatives and the European Union's Integrated Product Policy which seek to extend the responsibility of producers to the entire product life cycle.

The concept of green supply chain management (GSCM) has emerged from life cycle thinking and its application (Seuring and Muller, 2007). As Figure 1.4 shows, it includes environmental considerations in the total value chain from original source of raw materials, through the various companies involved in extracting and processing, manufacturing, distributing, consumption and disposal (Saunders, 1997).

The adoption of GSCM is very demanding as it requires, in addition to various elements of cleaner production and the implementation of EMS, the development and maintenance of close co-operative relations with external entities such as suppliers and retailers.

In recent years, the pressure for companies to be accountable for their environmental and social responsibilities has risen. This has led to the concept and practice of corporate social responsibility (CSR) whereby companies, on a voluntary basis, declare their commitment to consider the ethical consequences of their business activities and to take responsibilities for them beyond legal requirements.

Figure 1.4. Life cycle thinking

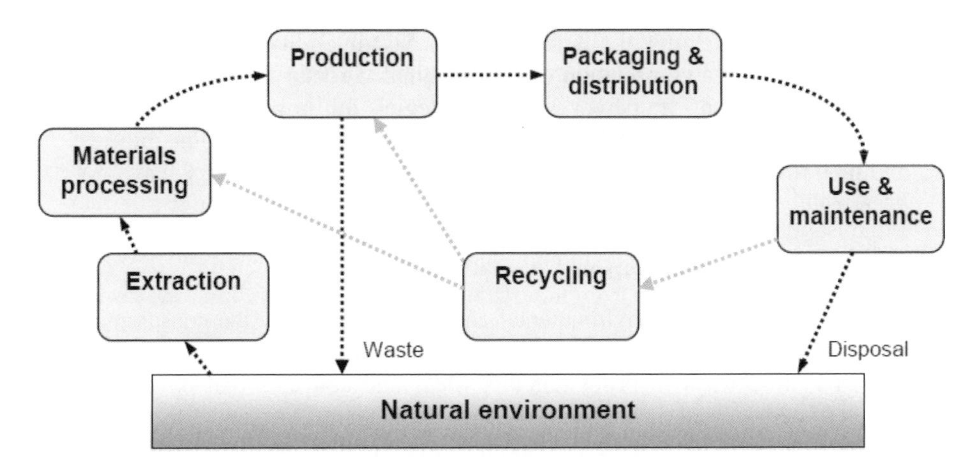

In recent years, CSR has emerged as a mainstream business issue, mostly owing to growing attention to social and environmental issues and rising demand for improved business ethics from governments, activists, the media, investors and the like (Porter and Kramer, 2006). CSR is primarily voluntary but some governments are exerting pressures on companies to improve their accountability, for example by requiring the disclosure of ethical, social and environmental risks in annual corporate reporting (*e.g.* France's new economic regulations of 2001).

Box 1.2. Corporate sustainability reporting

Public sustainability reporting on the environmental and social activities of companies and their supply chain provides a way for companies to inform stakeholders about their accomplishments and sustainable development targets. Reporting is typically voluntary but can be considered as a company's non-financial equivalent to its financial report.

Even though sustainability reporting has been mostly used as a communication tool, it is nevertheless widely recognised as an important mechanism for improving corporate environmental and social performance. A growing number of companies have also engaged in sustainability reporting because bank and investment managers increasingly look into what lies beyond the balance sheet. International initiatives such as the UN Global Compact and the UN Principles for Responsible Investment (PRI) are adding to the pressure on companies to report on their sustainability performance.

Today, several frameworks and guidelines on how and what to report exist. The Global Reporting Initiative's Sustainability Reporting Guidelines are becoming an internationally accepted standard (see Chapter 3).

Yet, while a growing number of companies now address CSR issues, they are often not clear on what exactly is involved and which concrete actions they take (Porter and Kramer, 2006). Sustainability reports (Box 1.2) also tend to offer a compilation of un-coordinated social and environmental activities. Coherent frameworks and strategies for how the company is addressing, or plans to address, its social and environmental responsibilities, and how these are linked to the company's core business strategy, have not been widely addressed (GRI and KPMG, 2008).

A new industrial revolution

To meet the global environmental challenges created by the consumption and production patterns established since the Industrial Revolution, there is a need to find ways to bring together ideas and concepts that have traditionally been viewed as trade-offs. In essence, there is a need for a "New Industrial Revolution" where economic wealth goes hand in hand with environmental and social sustainability. The increasingly blurred demarcation of manufacturing and services (Mont, 2002), or goods and services, can be seen as an early example of developments in this direction. Switching towards better environmental performance through reduced material flows has led to a more integrated approach to sustainable manufacturing, often referred to as a product-service system (PSS). PSS encourages companies to increase the re-use and remanufacturing of products. Taking this further, the need for virgin materials can be drastically reduced by adopting closed-loop production which maximises recycling of materials that already exist in the production system. Advanced solutions adopt an even more holistic view, such as industrial ecology in which the effluents of one producer's operations are used in another's production.

Product-service system (PSS)

Whereas traditional manufacturing focuses on the production and supply of goods to consumers, a PSS focuses on the delivery of consumer utility and product functionality. For example, when producing and supplying photocopiers to their consumers, a company based on the PSS model retains product ownership and supplies the photocopier as a function. In this way consumers purchase the copying service and not the product itself.

The PSS concept is widely discussed in sustainability-related articles but rarely in the mainstream business literature (Tukker *et al.*, 2006). In the latter, however, concepts such as "functional sales" and "servicising" have a similar meaning. In fact, the PSS approach has been applied in business-to-business contexts for many years. Since product ownership is not transferred from the producer to the consumer, the costs of product maintenance, retirement and replacement are internalised for the producer's profit maximisation objectives. As such, because the entire stock of manufactured goods is essentially "stored" by consumers, companies need not sell more products to maximise profits. Instead, they can reap profits by minimising material consumption and increasing product reuse, recycling and remanufacturing. This can result in far-reaching environmental benefits.

Product-use intensity is another environmental benefit that could be gained from PSS by sharing the same products among many consumers. Today, a car is parked rather than driven most of the time and an electric drill is typically used a few times a year. The PSS could lead to a radical reduction in the production of physical goods and thus to less consumption of materials and generation of waste. PSS also offers the opportunity to alleviate the pressure of realising profits in markets characterised by rapid changes in consumer preferences and in technological developments (Behrendt *et al.*, 2003).

The adoption and financial viability of PSS depends on the degree of change in economic, social and technological infrastructures as well as business models (Mont, 2002). From the perspective of manufacturing companies, for instance, PSS could imply a shift from the traditional point-of-sale business model to one centred on long-term service contracts. This would affect the organisational management and marketing of products. The major issue from consumers' perspective is product ownership. For the PSS model to function, consumers need to see products as leased rather than owned and shared rather than used. However, ownership of certain products is strongly entangled with consumers' identity and status (*e.g.* cars, luxury goods, houses) (Box 1.3).

Box 1.3. An application of product-service systems

InterfaceFLOR, an American producer of carpets, is offering carpet rotation and replacement services instead of selling carpets. This PSS is part of a broader initiative called "Mission Zero" through which the company aims to eliminate all forms of waste from its facilities by 2020, including carpets that are sent to landfill after usage. The company is using the rotation and replacement service as a model to take back old carpets for what they call "re-entry" – recycling materials that can be used for new carpets to decrease the use of virgin petroleum-based raw materials.

Source: InterfaceFLOR website, *www.interfaceflor.com*.

Closed-loop production

Closed-loop production is similar to life cycle thinking but distinguishes itself by "closing" the material resource cycle (Figure 1.5). This implies that all components that exist in the system are reused, remanufactured or recycled in some way. This entails a shift from traditional linear production methods to a circular and more systemic perspective in which products and processes are designed with "reincarnation" in mind. The need for virgin materials is eliminated, or drastically reduced, and waste is recycled into the system. Closed-loop production, therefore, constitutes advancing "cradle to grave" thinking towards "cradle to cradle" (McDonough and Braungart, 2002).

Figure 1.5. Closed-loop production system

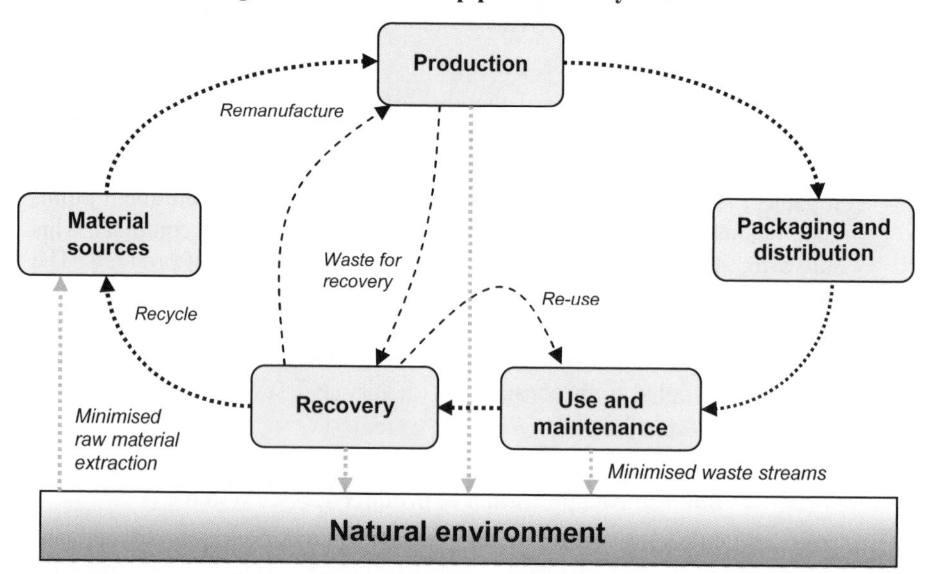

The development of closed-loop manufacturing requires a strong focus on the product design process. In addition to minimising the material and energy use needed to make and distribute products as well as the impacts from product use and disposal, the design process must also take into account means of recovering products and waste. For heavy machinery, for instance, vehicle design can be optimised not only by using the fewest possible harmful materials and aiming for the highest fuel efficiency, but also by designing the vehicles for disassembly/separation, cleaning, inspecting, repairing, replacing, a long life-time, and reassembling and "rebirth". By tapping into the large resource potential that exists in current waste, the need for virgin materials and waste disposal could be significantly reduced. PSS can facilitate business conditions for realising closed-loop production as an important building block for sustainable manufacturing (Behrendt *et al.*, 2003) (Box 1.4).

Box 1.4. Remanufacturing and PSS

Remanufacturing is a practice that can reduce environmental impacts while increasing revenue. Caterpillar, an American construction and mining equipment manufacturer, has embraced this idea as an integral part of its business model and has improved its environmental conduct by doing so. It established ongoing revenue opportunities for several generations of their product lines through new design strategies and collection mechanisms that maximise remanufacturing possibilities. Using financial incentives for customers to return equipment after the end of its life, the company is able to remanufacture components for a fraction of the original cost while keeping attractive profit margins even if the remanufactured products are sold at discount prices with the same warranties as new products.

Source: Gray and Charter (2006), *Remanufacturing and Product Design*, Centre for Sustainable Design, Farnham.

Industrial ecology

The extensive application of closed-loop production views and techniques across industries and society at large, *i.e.* beyond the boundary of a single company, is called industrial ecology. Industrial ecology, which stems from systems theory, views environmental ecology and uses natural eco-systems as a metaphor and model for better organising industrial production (Frosch and Gallopoulos, 1989). More specifically, industrial ecology considers the industrial production system as an interdependent part of the eco-system (Garner and Keoleian, 1995). That is, the industrial society must be understood not in isolation from its surrounding systems but in harmony with them (Jelinski *et al.*, 1992).

With respect to closed-loop production, industrial ecology might be viewed as "a system of systems", which ties several closed-loop production systems together by a circular flow of resources such that one system's effluents are used as another system's input, while also operating in harmony with the greater ecosystem. This means that industrial ecology not only relies on materials that can be recycled in the industrial production system, such as aluminium, but also on materials that are reusable in the natural environment, such as textiles that can serve as biodegradable garden mulch after life as an upholstery fabric. Mimicking eco-system terminology, these materials can be referred to as technical and biological nutrients (McDonough and Braungart, 2002). The development and implementation of such a system necessitates a multidisciplinary and multi-organisational approach in which stakeholders from various industrial sectors, areas of society and disciplines engage in intelligent and co-operative partnerships. Thus no company can become sustainable on its own.

At present, there is a considerable gap between theoretical approaches to industrial ecology and what is being implemented in a world in which the value chain of manufacturing companies is increasingly globalised. However, some applications of industrial ecology have been attempted through the establishment of "eco-industrial parks". These parks are comprised of a cluster of companies that seek to harness industrial symbioses through close co-operation with each other, and with the local community, by sharing resources to improve economic performance while minimising waste and pollution (Box 1.5). This idea is also promoted by the United Nations University (UNU) Zero Emissions Forum, which is establishing pilot eco-park projects as well as researching industrial synergies and sustainable transactions (Kuehr, 2007).

Box 1.5. An eco-industrial park in Denmark

One of the earliest and best-known eco-industrial parks is located in Kalundborg, Denmark. Rather than the result of a carefully planned process, the eco-park has developed gradually through co-operation among a number of neighbouring industrial companies. The main participating companies are a coal-fired power plant (Asnæsværket), a refinery (Statoil), a pharmaceutical and industrial enzyme plant (Novo Nordisk and Novozymes), a plasterboard factory (Gyproc), a soil remediation company (AS Bioteknisk Jordrens), and the municipality of Kalundborg through the town's heating facility.

The eco-park began when Gyproc located its facility in Kalundborg in 1970 to take advantage of the butane gas available from the Statoil refinery. At the same time this enabled Statoil to stop flaring the gas. Since then, the network has grown and today the participating companies are highly integrated. For instance, surplus heat from the power plant is used to heat about 4 500 private homes and water for fish farming, and fly ash is supplied for production of cement. Process sludge from fish farming and Novo Nordisk is supplied to nearby farms as fertiliser. Novo Nordisk also supplies farms with surplus yeast from insulin production for pig food. The Statoil refinery supplies pure liquid sulphur from its desulphurisation operations to a sulphuric acid producer (Kemira).

The exchanges above only describe a part of the material flow of the Kalundborg eco-park, which in total has been estimated to be around 2.9 million tonnes a year including fuel gases, sludge, fly ash, steam, water, sulphur and gypsum. This industrial symbiosis has served to reduce the environmental impacts of industrial production and led to significant economic savings. The participating companies are constantly co-operating to find new ways of improving the industrial symbiosis based on economic and environmental consciousness.

Source: Industrial Symbiosis Institute website *www.symbiosis.dk*;
Gibbs (2008), "Industrial Symbiosis and Eco-industrial Development: An Introduction", *Geography Compass*, Vol. 2, No. 4.

Summing up

To sum up, the thinking and practices surrounding sustainable manufacturing have evolved in several ways in the last decades, from the application of technology for the treatment of pollution at the end of the pipe through prevention of pollution to minimising inputs and outputs and substituting toxic materials. Recently, manufacturing companies have focused on solutions that integrate methods of minimising material and energy flows by changing products/services and production methods and revitalising disposed output as new resources for production.

Advances towards sustainable manufacturing have also been achieved through better management practices. Environmental strategies and management systems have allowed companies to better identify and monitor their environmental impacts and have facilitated improvements in environmental performance. Although such measures were initially limited to plant-specific production systems, they have evolved towards support for better environmental management throughout the life cycle of products and the value chain of companies.

Figure 1.6. The evolution of sustainable manufacturing concepts and practices

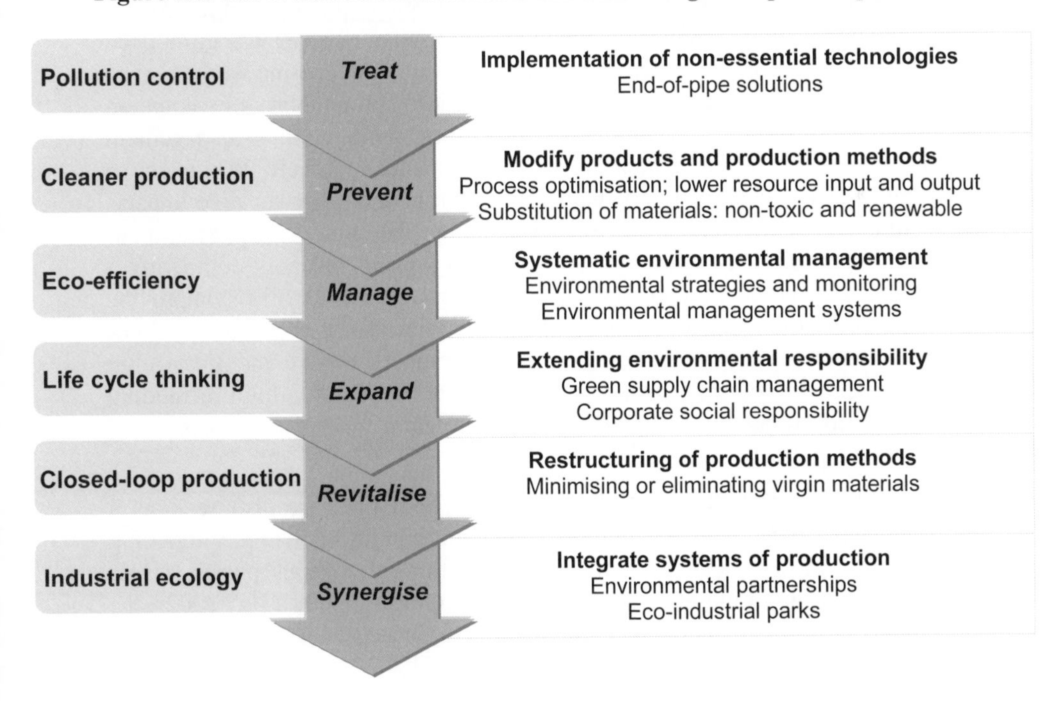

More integrated and systematic methods to improve sustainability performance in manufacturing industries have laid the foundation for the introduction of new business models such as PSS which could lead to significant environmental benefits. Furthermore, although still few in numbers, more efficient and intelligent ways of structuring production systems are being established, such as eco-industrial parks in which economic and environmental synergies between traditionally unrelated industrial producers are harnessed (Figure 1.6).

Understanding eco-innovation

In the last few years, many companies and consulting firms have started using eco-innovation or similar terms to present positive contributions by business to sustainable development through innovation and improvements in production processes and products/services. A few governments and the European Union (EU) are now promoting the concept as a way to meet sustainable development targets while keeping industry and the economy competitive.

In the EU, eco-innovation has been considered to support the wider objectives of its Lisbon Strategy for competitiveness and economic growth. In 2004, the Environmental Technology Action Plan (ETAP) was introduced to promote the development and implementation of eco-innovation.[4] The ETAP defines eco-innovation as "the production, assimilation or exploitation of a novelty in products, production processes, services or in management and business methods, which aims, throughout its life cycle, to prevent or substantially reduce environmental risk, pollution and other negative impacts of resource use (including energy)". The action plan provides a general roadmap for promoting environmental technologies and business competitiveness by focusing on bridging the gap between research and markets, improving market conditions for environmental technologies, and acting globally. Eco-innovation now forms part of the EU's Competitiveness and Innovation Framework Programme 2007-13, which offered EUR 28 million in funding in 2008 to stimulate the uptake of environmental products, processes and services especially among SMEs.

In the United States, environmental technologies are also seen as a promising means of improving environmental conditions without impeding economic growth, and are being promoted through various public-private partnership programmes and tax credits (OECD, 2008). In 2002, the Environmental Protection Agency laid out a strategy for achieving better environmental results through innovation (EPA, 2002). Based on this strategy, it set up the National Center for Environmental Innovation and is promoting the research, development and demonstration of technologies that contribute to sustainable development in partnership with state governments, businesses and communities.

While the promotion of eco-innovation so far has focused mainly on the development and application of environmental technologies, there is an increasing emphasis on going beyond these. This reflects the growing understanding of and research on the non-technological aspects of innovation, such as organisational innovation and marketing innovation, as defined in the latest version of the OECD's *Oslo Manual* (OECD and Eurostat, 2005). It also reflects the fact that eco-innovation's focus on sustainable development demands broad structural changes in society.

In Japan, the government's Industrial Science Technology Policy Committee introduced the term eco-innovation in 2007 as an overarching concept which provides direction and a vision for the societal and technological changes needed to achieve sustainable development. The committee considers that the current pattern of economic growth achieved through "functionality-oriented, supplier-led mass consumption" is approaching its limit owing to constraints on the environment, resources and energy. As Japan's people have been highly satisfied in material terms, it argues that economic growth in the 21st century can be pursued by appealing to people's *kansei* (sensitivity). This would also require the establishment of a new socio-industrial structure in which environmental conservation and economic growth are fused. In short, the committee defines eco-innovation as "a new field of techno-social innovations [that] focuses less on products' functions and more on [the] environment and people". In more concrete terms, the committee proposes promoting the construction of "zero emission-based" infrastructures in energy supply, transport and town development, as well as sustainable lifestyles by selling services instead of products and by promoting environmental and *kansei* values (METI, 2007).

While overall aims for promoting eco-innovation seem to have in common the parallel pursuit of economic and environmental sustainability, there is some diversity in the application of the concept. To improve the conceptual understanding of eco-innovation and to facilitate the construction of an analytical framework that combines eco-innovation with sustainable manufacturing, this section attempts to draw together a conceptual and typological overview of eco-innovation and the different areas to which the concept can be applied for diverse types of businesses.

A conceptual overview

The term eco-innovation seems to have first appeared in *Driving Eco-Innovation,* a book by Claude Fussler and Peter James in 1996. The authors defined the concept as "new products and processes that provide customer and business value while significantly decreasing environmental impacts". Under the overarching concept of sustainable development the meaning of eco-innovation has come to include social and institutional aspects. Although some strands in the literature attempt to discern and highlight differences between concepts such as "eco-innovation", "environmental innovation", "innovation for sustainable development" and "sustainable innovation", they are mostly used interchangeably (Charter and Clark, 2007). This chapter primarily uses the term eco-innovation but makes no distinction with the related concepts.[5]

Eco-innovation is closely related to the conventional understanding of innovation which, according to the *Oslo Manual* (OECD and Eurostat, 2005), can be described as the implementation of new, or significantly improved, products (goods or services), or processes, marketing methods, or organisational methods in business practices, workplace organisation or external relations. It is distinct from invention, which refers to the phase in which the idea behind the innovation is conceived. It is also distinct from the dissemination of the innovation. Combined, however, invention, innovation and dissemination constitute what is referred to as the innovation process. This process should also be applicable to eco-innovation.

Eco-innovation can, however, be distinguished from conventional innovation in two significant ways. First, it is not an open-ended concept as it represents innovation which explicitly emphasises the reduction of environmental impacts, whether intended or not. Second, eco-innovation is not limited to innovation in products, processes, marketing methods and organisational methods, but also includes innovation in social and institutional structures (Rennings, 2000). This reflects the fact that the scope of eco-innovation can extend beyond the conventional organisational boundaries of the innovating company to encompass the broader societal sphere. It thus involves changes in social norms, cultural values and institutional structures – in partnership with stakeholders such as competitors, companies in the supply chain, those from other sectors, governments, retailers and consumers – to leverage more environmental benefits from the innovation.

Based on the *Oslo Manual* and drawing from other sources (*e.g.* METI, 2007; Reid and Miedzinski, 2008; MERIT *et al.*, 2008),[6] eco-innovation can be described as "the implementation of new, or significantly improved, products (goods and services), processes, marketing methods, organisational structures and institutional arrangements which, with or without intent, lead to environmental improvements compared to relevant alternatives". On this interpretation, innovation and eco-innovation are distinguished from relevant alternatives solely by their environmental effects. The definition therefore only provides a weak conceptual demarcation of innovation and eco-innovation and should only be seen as a starting point for analysis of eco-innovation. To facilitate the analysis of different business activities aimed at eco-innovation, the concept and its typology are further elaborated below.

A typology

Inspired by existing innovation and eco-innovation literature (*e.g.* OECD and Eurostat, 2005; Charter and Clark, 2007; Reid and Miedzinski, 2008), it is proposed that an eco-innovation can be understood on the basis of three key axes: its target, its mechanism and its impact:

- **Target** refers to the basic focus of eco-innovation. Building upon the typology of the *Oslo Manual*, the target of an eco-innovation can be categorised under: *i)* products (both goods and services); *ii)* processes, such as a production method or procedure; *iii)* marketing methods, referring to the promotion and pricing of products, and other market-oriented strategies; *iv)* organisations, such as the structure of management and the distribution of responsibilities; and *v)* institutions, which include broader societal areas beyond a single company's control such as broader institutional arrangements as well as social norms and cultural values.
- **Mechanism** relates to the method by which the change in the eco-innovation target takes place or is introduced. It is also associated with the underlying nature of the eco-innovation, *i.e.* whether the change is technological or non-technological in nature. Four basic mechanisms are identified: *i)* modification, such as small, progressive product and process adjustments; *ii)* redesign, referring to significant changes in existing products, processes, organisational structures, etc.; *iii)* alternatives, such as the introduction of goods and services that can fulfil the same functional needs and operate as substitutes for other products; and *iv)* creation, comprising the design and introduction of entirely new products, processes, procedures, and organisational and institutional settings.
- **Impact** refers to the eco-innovation's effect on environmental conditions, across its life cycle or some other scope. The impact depends on the combination of the innovation's target and mechanism, here referred to as the innovation's design, and can be illustrated across a continuous range starting from incremental environmental improvements to the complete elimination of environmental harm. For particularly well-defined areas, it can be related to the concept of "Factor" which is used to describe technological performance with respect to energy and resource efficiency (Weizsacker *et al.*, 1998).

A Factor 2 improvement in CO_2 emissions, for example, denotes a 50% reduction, everything else being equal.

Based on this typology, companies can design and analyse their eco-innovative initiatives and strategies with respect to specific areas (targets), the type of progress that is being made (mechanisms), and the resulting effects (impacts). While this approach can be applied to eco-innovative initiatives across all targets and mechanisms, it is generally possible to distinguish the underlying nature of change with respect to eco-innovation in products and processes from that in marketing methods, organisations and institutions. Eco-innovation in products and processes, for instance, is typically considered more closely related to technological advances regardless of the eco-innovation's basic mechanism. For marketing methods and organisational structures, on the other hand, eco-innovative mechanisms tend to be associated with non-technological changes (OECD, 2007). This notion extends to changes in institutional arrangements. These differences, along with the impact of eco-innovation, are further illustrated below.

Eco-innovation in products and processes

Advances in products and processes, which tend to rely on technological change, cover a broad range of tangible objects that can improve environmental conditions and might therefore be referred to as technological eco-innovations. Examples include computer chips that are faster but consume less energy, cars that are more fuel-efficient, and production methods that use fewer resources. Generally, they are also curative or preventive in nature.

Curative eco-innovative technologies are equivalent to the end-of-pipe technologies described above, because they seek to reduce or eliminate contaminants that have already been produced. Preventive eco-innovative technologies, on the other hand, aim to reduce or eliminate the source of the pollutants. These technologies are thus related to cleaner production techniques but may be unintended results of efforts to improve general business profitability.

Both curative and preventive eco-innovative products and processes can tackle environmental challenges. Yet, from a broader sustainability perspective, they should only be seen as part of the solution (Brown *et al.*, 2000). Moreover, if they are not tested with a view to their potential adverse effects, some may even create new environmental hazards and problems (Reid and Miedzinski, 2008) (Box 1.6).

Box 1.6. The rise and fall of CFC gases

Chlorofluorocarbon (CFC) gases were developed in the 1930s to replace hazardous materials such as sulphur dioxide and ammonia. Owing to their non-toxic, non-flammable and non-corrosive properties, and being both inexpensive and efficient, they were long considered to be an ideal refrigerant. The use of CFCs increased rapidly after their market introduction not only in air conditioning and refrigeration equipment but also in a large range of industrial applications.

In the 1970s, however, it was found that CFC gases have an ozone-depletion effect. Large reductions in the ozone layer, particularly over Antarctica, were reported in the mid-1980s and concerns arose about the increased likelihood of skin cancer. This eventually led to the ban of CFC gases under an international agreement when the Montreal Protocol on Substances that Deplete the Ozone Layer entered into force in 1989.

Source: World Meteorological Organization (WMO) and United Nations Environment Programme (UNEP) (1998), *Scientific Assessment of Ozone Depletion: 1998*, WMO Ozone Report No. 44, WMO, Geneva; WMO and UNEP (2006), *Scientific Assessment of Ozone Depletion: 2006*, WMO, Geneva.

Eco-innovation in marketing, organisations and institutions

Contrary to products and processes, eco-innovation in marketing methods, organisational structures and institutional arrangements tends to rely on non-technological mechanisms. Such changes constitute a relatively new area in the innovation literature and were only covered in the third and latest revision of the *Oslo Manual* in 2005 by the introduction of innovation in marketing methods and organisational structures.

Eco-innovation in marketing includes new ways of integrating environmental aspects in communication and sales strategies. Eco-innovative marketing concerns the company's orientation towards customers and can play a significant role in leveraging environmental benefits by influencing them. For instance, the company can improve general product and company appeal in connection with the development and/or sale of eco-efficient products through better market research, direct contact with consumers, and marketing practices that appeal to environmentally aware consumers. Eco-innovation in marketing may also include new business models that change the way products are priced, offered and promoted such as the adoption of PSS.

Organisational eco-innovation includes the introduction of new management methods such as EMSs and corporate environmental strategies. While these areas concern general environmental business practices, organisational eco-innovation can also take place through changes in the company workplace, such as the centralisation or decentralisation of environmental responsibilities and decision-making powers or the establishment of training programmes for employees designed to improve environmental awareness and performance. Organisational eco-innovation also includes changes in how companies organise their relations with other firms and public institutions, such as the adoption of GSCM and the participation in public-private partnerships for environmental research and projects.

Although institutional innovation is not covered by the *Oslo Manual*, the literature on conventional innovation emphasises the importance of co-evolving social and institutional changes in connection with, but as a separate part of, the innovation process (Grubb, 2004; Reid and Miedzinski, 2008). In the context of sustainability, however, a small but growing body of literature argues that changes in social norms, cultural values and institutional structures can be considered eco-innovation in themselves or constitute integral parts of an eco-innovation (Rennings, 2000). This view is gaining ground from a policy perspective. In Japan for instance, eco-innovation is increasingly viewed as a field of techno-social innovations that not only can improve environmental conditions but also satisfy subjective values (METI, 2007).

The concept of institutions generally covers a wide range, from social norms and cultural values to codified laws, rules and regulations, and from loosely established social arrangements to deliberately created institutional frameworks. In some cases institutions are seen as exogenous and outside the domain of market transactions; in others they are seen as endogenous (van de Ven and Hargrave, 2002; Aoki, 2007). This study distinguishes between informal institutions such as social norms and cultural values, which tend to be endogenous, and formal institutions such as codified laws, regulations, and formal institutional frameworks and arrangements, which tend to be based on policy and economic decisions.

Eco-innovation in informal institutions refers to changes in value patterns, beliefs, knowledge, norms, etc., that lead to improvements in environmental conditions through social behaviour and practices. For instance, this would include shifts in the choice of transport modes, *i.e.* from personal automobiles or flights to trains, buses or bicycles because of users' higher environmental awareness or education. It may also include the growth of self-help health groups, community action for cleaning up the surrounding environment, organic food movements, etc.

Formal institutional eco-innovation refers to structural changes that redefine roles and relations across a number of independent entities. It typically relies on legal enforcement, international agreements, or voluntary but formal multi-stakeholder arrangements. Institutional eco-innovative solutions may range from agencies to administer clean local water supplies, financial platforms for funding the development of environmental technologies, and the establishment of eco-labelling schemes and environmental reporting frameworks to new regimes of global governance such as the establishment of an institution with responsibility for global climate and biodiversity issues (Rennings, 2000). In terms of sustainable manufacturing, the establishment of eco-industrial parks, where resource sharing is optimised across seemingly unrelated industrial producers can be considered an example of formal institutional eco-innovation.

Impacts of eco-innovation

The environmental impact of an eco-innovation stems from the interplay between the innovation's design (target and mechanism) and the socio-technical environment in which the innovation is introduced. From an analytical perspective, the assessment of this impact is very important because it determines whether or not the eco-innovation can in fact be classified as such. Also, from a practical point of view, it is important to show that the eco-innovation improves overall environmental conditions. However, the impact assessment of eco-innovation requires extensive knowledge and understanding of the innovation and its contextual relationships.

For example, rather simple adjustments that are not intended to improve environmental performance can have significant environmental benefits. These may occur as a result of an unexpected interaction with other factors and occur through indirect systemic changes. An illustrative example is the provision of power outlets and wireless Internet connections in trains. While these adjustments require extra resources and consume additional energy, thus leading directly to a decline in environmental performance, the overall environmental impact could more than offset this negative effect if the new facilities, through "green marketing", attracted business travellers who otherwise would travel by air or automobiles.

Hence, eco-innovation assessments must consider the eco-innovation's life cycle at several levels (Reid and Miedzinski, 2008), including the behavioural and systemic consequences of the innovation's application and/or usage. These can be categorised according to the innovation's characteristics at the micro level, referring to companies and individuals; at the meso level, including supply chains, sectoral structures, local perspectives, etc.; and at the macro level, referring to countries, economic blocs and the global economy. A problem in this regard is the lack of recognised methodological approaches, in part because eco-innovation remains a relatively unrecognised field in innovation policy and general policy frameworks (MERIT *et al.*, 2008).

Summing up

To sum up, eco-innovation can be categorised according to its target (products, processes, marketing methods, organisational structures and institutions); its mechanism (modification, redesign, alternatives and creation); and its environmental impact. The target of the eco-innovation can generally be associated with its technological or non-technological nature: eco-innovation in products and processes tends to rely heavily on technological development, and eco-innovation in marketing, organisations and institutions relies more on non-technological changes. Potential environmental impacts stem from the eco-innovation's target and mechanism and their interplay with the innovation's socio-technical context. Given a specific target, the magnitude of the environmental impact nevertheless tends to follow the eco-innovation's mechanism: modifications generally lead to lower potential environmental benefit than creations. Figure 1.7 sketches an overview of eco-innovation and its typology.

Figure 1.7. The typology of eco-innovation

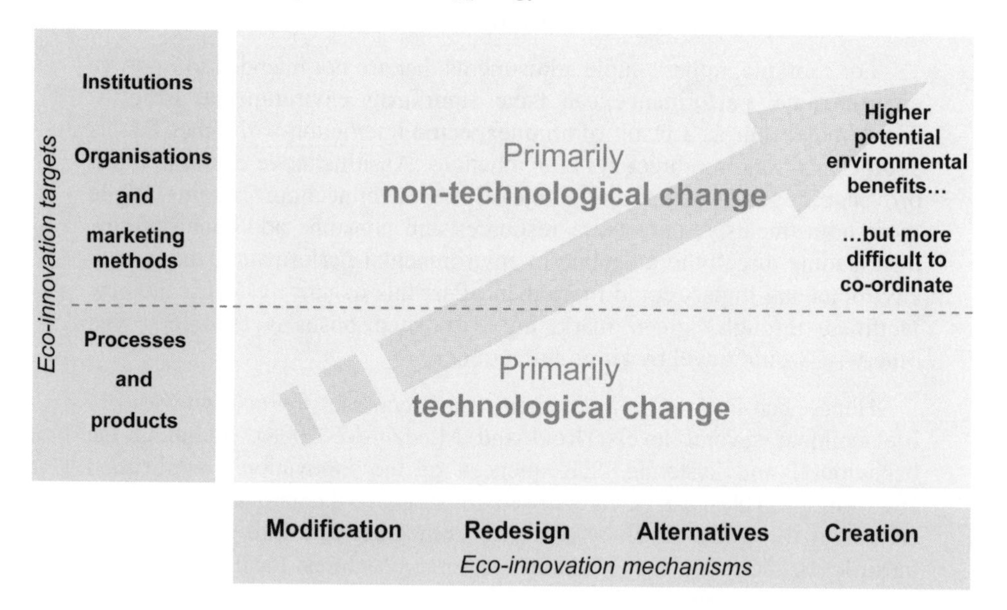

So far, the primary focus of eco-innovation, as of conventional innovation, has been the development and application of different technologies, but recent evidence suggests that non-technological changes are becoming more important (Reid and Miedzinski, 2008). It is also important for eco-innovative solutions to go beyond products, processes, marketing methods and organisational structures, and start to tap into areas relating to social norms, cultural values and formal institutional structures. This is particularly important because the greatest potential for system-wide environmental improvements is typically associated with the development of new social structures and interactions, including changes in value patterns and behaviour, rather than in incremental technological advances.

Eco-innovation as a driver of sustainable manufacturing

There are clearly many conceptual overlaps between eco-innovation and sustainable manufacturing. Pollution control, for instance, can be related to the modification of products and processes; cleaner production initiatives are often associated with the implementation of more integrated changes such as redesign of products and production methods. Eco-efficiency and life cycle thinking are related to eco-design of products and processes, as well as the adoption of EMSs and GSCM. Closed-loop production may refer to alternative business models such as the adoption of PSS, while industrial ecology can generally be associated with the creation of entirely new production structures.

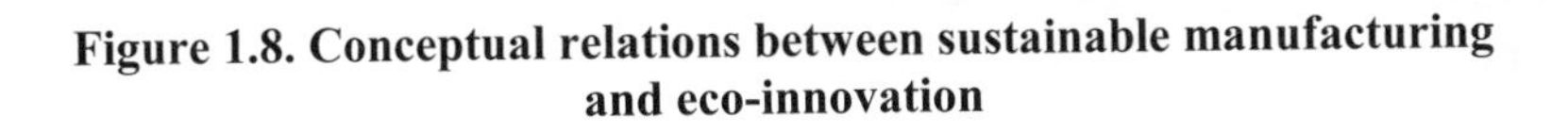

Figure 1.8. Conceptual relations between sustainable manufacturing and eco-innovation

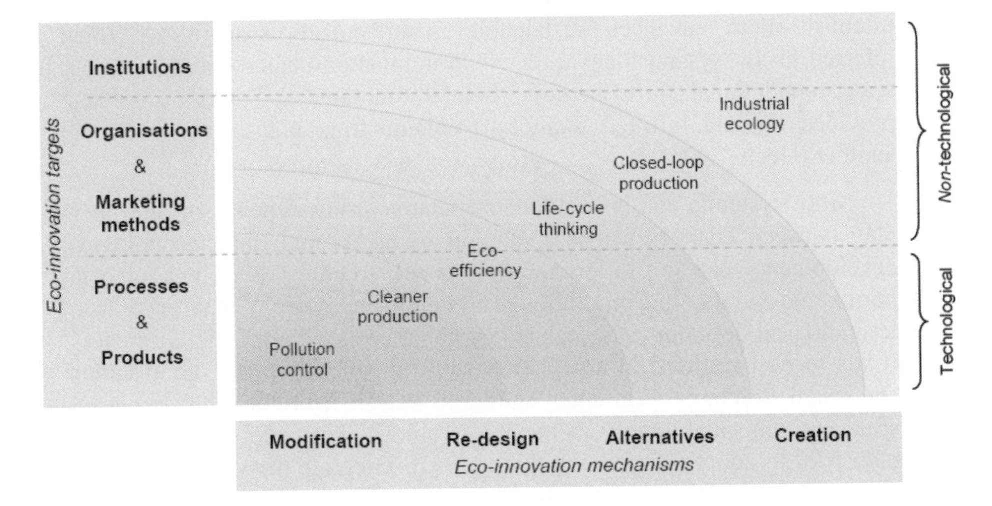

Using Figure 1.7 as a basis for understanding eco-innovation, Figure 1.8 attempts to give a simple illustration of the general conceptual relations and overlaps that exist between the concepts of sustainable manufacturing and eco-innovation. The evolutionary steps of sustainable manufacturing are depicted in terms of their primary association with eco-innovation, *i.e.* with innovation targets on the left, and mechanisms at the bottom. The underlying nature of eco-innovation (technological or non-technological) is depicted on the right. The "waves" spreading towards the upper right-hand corner of the figure indicate the path dependencies of different sustainable manufacturing concepts.

In the medium to long term, the most potentially significant environmental improvements from eco-innovation in manufacturing industries are associated with more advanced sustainable manufacturing initiatives such as the establishment of eco-industrial parks and the like. However, these can generally only be realised through a combination of a broader range of innovation targets and mechanisms; hence those initiatives cover the bigger area of the figure. It is not enough, for instance, simply to locate manufacturing plants with symbiotic relationships close together if no technology or procedure for exchanging resources exists. Process modification, product design, business model alternatives and the creation of new methods, procedures and arrangements should go hand in hand and must evolve together to leverage the economic and environmental benefits from such initiatives. This also means that as sustainable manufacturing initiatives advance, the nature of the eco-innovation process becomes increasingly complex and more difficult to co-ordinate.

The co-evolutionary eco-innovation processes that are necessary to establish more advanced sustainable manufacturing systems are often referred to as "system innovation" – an innovation characterised by large-scale foundational shifts in how societal functions and needs are being provided for and fulfilled, such as a change from one energy source to another (Geels, 2005).

More systemic eco-innovation in manufacturing depends on the interplay between changes across a number of areas, including technological developments, changes in formal institutional structures as well as in social norms and values. Indeed, although systemic innovations may arise from technological developments, technology alone cannot make large differences. It has to be harnessed in association with human enterprise, organisations and social structures. While this highlights the difficulty of achieving large-scale environmental improvements, it also hints at the need for manufacturing industries to adopt an approach that seeks to integrate the various elements of the eco-innovation process, in such a way that the interplay of changes leverages environmental benefits (Box 1.7 gives advanced examples).

Box 1.7. Examples of eco-innovative solutions

The BMW Group, which has been developing hydrogen engine technologies for more than 25 years, has recently unveiled a new "mono-fuel" internal combustion engine. The engine is introduced in the new mono-fuel Hydrogen 7 saloon, which was first displayed at the SAE World Congress in Detroit in 2008. Initial testing of the exhaust from the car's near-zero-emissions engine shows that the air is cleaner in components such as non-methane organic gases (NMOGs) and carbon monoxide (CO) than the air coming in as the engine absorbs and burns ambient air pollutants.

McDonough Braungart Design Chemistry (MBDC), which was established in 1995 to advance the "New Industrial Revolution" and the realisation of the "cradle-to-cradle" thinking, developed an ice cream package for Unilever based on eco-innovative thinking. The packaging consists of polymers, which take the form of a film in its frozen state but degrades to a liquid over a couple of hours when exposed to room temperature. The polymer packaging also includes seeds for rare plants. This essentially makes littering a way to improve biodiversity. It also demonstrates a radical conceptual change as waste literally creates potential new life.

Source: Wired (2008), "BMW Hydrogen 7 Mono-Fuel Eats Smog for Breakfast", 16 April; UNIDO (2002), "The New Industrial Revolution: Michel Braungart at Venice II", *UNIDO Scope Weekly News*, 20-26 October.

From an eco-innovation perspective, manufacturing industries have typically been more concerned with the modification and redesign of existing products, procedures and organisational structures than engaging in the creation of new and alternative solutions. The current focus and application of eco-innovative efforts in manufacturing industries have therefore been relatively narrow and limited to technical advances. This does not imply that environmental performance is not improving, but it can affect views of eco-innovative solutions and how they are developed and applied to manufacturing. It may also explain why the potentially transformative power of eco-innovation has remained largely peripheral in most corporate sustainability initiatives (Charter and Clark, 2007).

To conclude, eco-innovation plays a key role for driving manufacturing industries towards sustainable production. Every shift in environmental initiatives – from traditional pollution control to cleaner production initiatives and the establishment of eco-industrial parks – can be characterised as shifts facilitated by eco-innovation. The concept of eco-innovation can help companies and governments to consider and make these shifts through technological advances, changes in management tools, social acceptance of new products and procedures, as well as changes in institutional frameworks for facilitating progressive change.

Conclusions

The concept of sustainable development has been gaining attention in recent years and the topic has risen to the top of the international political agenda, particularly owing to concerns over climate change. Growing media coverage of environmental issues and rising public awareness have further increased the pressure for manufacturing industries to take responsibility by adopting more advanced and integrated responses to environmental concerns.

This has led to a substantial expansion of ways of applying sustainable development to production in general and to the establishment of a range of tools and management philosophies on sustainable business practices. In terms of sustainable manufacturing, this has involved a movement towards the application of technological solutions that enable the substitution of toxic materials by non-toxic alternatives and the reduction of material consumption and waste. With rising pressures on companies to take environmental responsibility beyond their organisational boundaries, many manufacturing companies have also adopted life cycle perspectives for their operations and are increasingly involved in green supply chain management. In recent years, the concept of a circular manufacturing process has gained ground and new business models, such as product-service system, which facilitate the move towards closed-loop production systems, have emerged. Many sustainable manufacturing initiatives, however, have primarily focused on the development and application of environmental technologies. While they have improved general environmental performance, environmental gains have mostly been incremental and in many cases have been outweighed by rising volumes of production and consumption (OECD, 2001).

To meet the growing environmental challenges, much attention has been paid to innovation as a way of developing sustainable solutions, also known as eco-innovation. This concept is gaining ground in industry and among policy makers as a way to facilitate the more radical and systemic improvements in corporate environmental performance that are increasingly needed. This has led to understanding eco-innovation in the sense that solutions concern not only technological developments but also non-technological changes such as those in consumer behaviour, social norms, cultural values, and formal institutional frameworks. Changes across all these areas, however, cannot be achieved by a single company (Jorna *et al.*, 2006; Reid and Miedzinski, 2008).

The concepts of sustainable manufacturing and eco-innovation are closely related, but not identical. Earlier and more traditional sustainable manufacturing initiatives, for instance, tend to take the form of adjustments to products and processes, marketing methods and organisational structures. Later and more advanced sustainable business practices, on the other hand, are related to the creation of new products and processes, alternative business models, and circular production systems in which discarded goods can be reutilised as new material inputs and seemingly unrelated industrial processes can be connected, with large environmental gains.

Eco-innovation can thus be understood as a driving force for moving manufacturing industries towards sustainable production. The application of the eco-innovation concept can offer a promising way to move industrial production towards true sustainability. However, it requires manufacturing industries to integrate and apply the concept in a more holistic way. It entails a deliberate re-examination of each phase of the production system in order to identify areas for applying potential eco-innovative solutions, including the development of new institutional arrangements such as knowledge networks and partnerships that can function as co-creative processes.

Notes

1. The US Department of Commerce (DOC) has recently defined sustainable manufacturing for the purposes of its Sustainable Manufacturing Initiative. It states that sustainable manufacturing is "the creation of manufactured products that use processes that minimize negative environmental impacts, conserve energy and natural resources, are safe for employees, communities, and consumers and are economically sound." See the DOC's Sustainable Manufacturing Initiative and Public-Private Dialogue website: *www.trade.gov/competitiveness/sustainablemanufacturing/how_doc_defines_SM.asp).*

2. In 1992, the UNCED concluded that "the major cause of the continued deterioration of the global environment is the unsustainable patterns of consumption and production, particularly in industrialized countries, which is a matter of grave concern, aggravating poverty and imbalances". This statement was put forward, particularly to Western countries, as a challenge to change current consumption and production patterns, backed by a global plan for action known as Agenda 21.

3. To address the difficulties in environmental performance measurement, the ISO issued the ISO 14031 standard in 1999 which contains guidance on the design and use of environmental performance evaluation in alignment with the ISO 14001 EMS standard.

4. The ETAP is actively seeking to consolidate an EU-wide market for environmental technologies. A core area is the development of an environmental technology verification (ETV) system that can help to accelerate market acceptance of key innovative technologies by providing accurate and verified information on technology performance. The European Commission is working closely with the United States and Canada where ETV systems have already been implemented.

5. Japan's eco-innovation concept aims at higher satisfaction of human needs and higher quality of life as well as environmental protection. In this publication, the concept of eco-innovation is only described in terms of its environmental aspects. However, the inclusion of social aspects can be considered by simple extension of the application areas and impacts of eco-innovation.

6. For example, the EU-funded Measuring Eco-Innovation (MEI) project proposes that eco-innovation be defined as "the production, assimilation or exploitation of a product, production process, service or management or business method that is novel to the organization (developing or adopting it) and which results, throughout its life cycle, in a reduction of environmental risk, pollution and other negative impacts of resources use (including energy use) compared to relevant alternatives" (MERIT *et al.*, 2008).

References

Aoki, M. (2007), "Endogenizing Institutions and Institutional Change", *Journal of Institutional Economics*, Vol. 3, No. 1, pp. 1-31.

Ashford, N.A. (1994), *Government Strategies and Policies for Cleaner Production*, UNEP, Nairobi.

Behrendt, S. *et al.* (2003), *Eco-Service Development: Reinventing Supply and Demand in the European Union,* Greenleaf Publishing, Sheffield.

Benyus, J.M. (1997), *Biomimicry: Innovation Inspired by Nature*, William Morrow & Co., New York.

Braungart, M. (2002), "The New Industrial Revolution", paper presented at the UNIDO, Venice II - Updating and Fleshing Out the Development Agenda, Venice.

Brown, D. *et al.* (2000), *Building a Better Future: Innovation, Technology and Sustainable Development*, WBCSD, Geneva.

Charter, M. and T. Clark (2007), *Sustainable Innovation: Key Conclusions from sustainable Innovation Conferences 2003-2006 Organised by The Centre for Sustainable Design,* Centre for Sustainable Design, Farnham.

European Commission (EC) (2005), *Doing More with Less: Green Paper on Energy Efficiency,* Office for Official Publications of the European Communities, Luxembourg.

Environmental Protection Agency, United States (EPA) (2002), *Innovating for Better Environmental Results: A Strategy to Guide the Next Generation of Innovation at EPA*, EPA, Washington, DC.

Frondel, M., J. Horbach and K. Rennings (2007), "End-of-Pipe or Cleaner Production?: An empirical Comparison of environmental Innovation Decisions across OECD Countries", in N. Johnstone (ed.), *Environmental Policy and Corporate Behaviour*, Edward Elgar, Cheltenham.

Frosch, R. and Y. Gallopoulos (1989), "Strategies for Manufacturing", *Scientific American*, Vol. 261, pp. 144-152.

Garner, A. and G.A. Keoleian (1995), *Industrial Ecology: An Introduction,* National Pollution Prevention Center for Higher Education, University of Michigan, Ann Arbor, MI.

Geels, F.W. (2005), *Technological Transitions and System Innovations: A Co-evolutionary and Socio-technical Analysis*, Edward Elgar, Cheltenham.

Gibbs, D. (2008), "Industrial Symbiosis and Eco-industrial Development: An Introduction", *Geography Compass*, Vol. 2, No. 4, pp. 1138-1154.

Gray, C. and M. Charter (2006), *Remanufacturing and Product Design*, Centre for Sustainable Design, Farnham.

Global Reporting Initiative (GRI) and KPMG (2008), *Reporting the Business Implications of Climate Change in Sustainability Reports*, GRI and KPMG, Amsterdam.

Grubb, M. (2004), "Technology Innovation and Climate Change Policy: An Overview of Issues and Options", *Keio Economic Studies*, Vol. 41, No. 2, pp. 103-132.

International Energy Agency (IEA) (2007), *Tracking Industrial Energy Efficiency and CO_2 Emissions*, OECD/IEA, Paris.

International Organization for Standardization (ISO) (2004), *ISO 14001:2004 Environmental Management Systems: Requirements with Guidance for Use*, ISO, Geneva.

International Union for Conservation of Nature (IUCN) (1980), *World Conservation Strategy: Living Resource Conservation for Sustainable Development*, IUCN, Geneva.

Jelinski, L.W. *et al.* (1992), "Industrial ecology: concepts and approaches", *Proceedings of the National Academy of Sciences*, Vol. 89, No. 3, pp. 793-797.

Johnstone, N. *et al.* (2007), "'Many a Slip 'twixt the Cup and the Lip': Direct and Indirect Public Policy Incentives to Improve Corporate Environmental Performance", in N. Johnstone (ed.), *Environmental Policy and Corporate Behaviour*, Edward Elgar, Cheltenham.

Jorna, R.J. *et al.* (2006), *Sustainable Innovation: The Organisational, Human and Knowledge Dimension*, Greenleaf Publishing, Cheltenham.

Kuehr, R. (2007), "Towards a Sustainable Society: United Nations University's Zero Emissions Approach", *Journal of Cleaner Production*, Vol. 15, No. 13-14, pp. 1198-1204.

Kurzinger, E. (2004), "Capacity Building for Profitable Environmental Management", *Journal of Cleaner Production*, Vol. 12, No. 3, pp. 237-248.

Maastricht Economic and Social Research and Training Centre on Innovation and Technology (MERIT) *et al.* (2008), *MEI Project about Measuring Eco-Innovation: Final Report*, under the EU's 6th Framework Programme, MERIT, Maastricht.

Maxwell, D., W. Sheate and R. van der Vorst (2006), "Functional and Systems Aspects of the Sustainable Product and Service Development Approach for Industry", *Journal of Cleaner Production*, Vol. 14, No. 17, pp. 1466-1479.

McDonough, W. and M. Braungart (2002), *Cradle to Cradle: Remaking the Way We Make Things* (1st ed.), North Point Press, New York.

Ministry of Economy, Trade and Industry, Japan (METI) (2007), *The Key to Innovation Creation and the Promotion of Eco-Innovation*, report by the Industrial Science Technology Policy Committee of the Industrial Structure Council, METI, Tokyo.

Mont, O. (2002), "Clarifying the Concept of Product-Service System", *Journal of Cleaner Production*, Vol. 10, No. 3, pp. 237-245.

Nasr, N. and M. Thurston (2006), *Remanufacturing: A Key Enabler to Sustainable Product Systems,* Rochester Institute of Technology, Rochester, NY.

OECD (2001), *Environmental Outlook 2001*, OECD, Paris.

OECD (2007), *Science, Technology and Industry Scoreboard 2007: Innovation, and Performance in the Global Economy*, OECD, Paris.

OECD (2008), *Environmental Innovation and Global Markets*, report for the Working Party on Global and Structural Policies, Environment Policy Committee, OECD, Paris, *www.olis.oecd.org/olis/2007doc.nsf/linkto/env-epoc-gsp(2007)2-final.*

OECD and Statistical Office of the European Communities (Eurostat) (2005), *Oslo Manual: Guidelines for Collecting and Interpreting Innovation Data* (3rd ed.), OECD, Paris.

Perotto, E. *et al.* (2008), "Environmental Performance, Indicators and Measurement Uncertainty in EMS Context: A Case Study", *Journal of Cleaner Production*, Vol. 16, No. 4, pp. 517-530.

Porter, M. E. and M. R. Kramer (2006), "Strategy and Society: The Link between Competitive Advantage and Corporate Social Responsibility", *Harvard Business Review*, Reprint No. R0612D.

Porter, M.E. and C. van der Linde (1995), "Green and Competitive: Ending the Stalemate", *Harvard Business Review*, Reprint No. 95507.

Reid, A. and M. Miedzinski (2008), *Eco-innovation: Final Report for Sectoral Innovation Watch,* Technopolis Group, Brighton.

Rennings, K. (2000), "Redefining Innovation: Eco-innovation Research and the Contribution from Ecological Economics", *Journal of Ecological Economics*, Vol. 32, pp. 319-332.

Roy, R. (2000), "Sustainable product-Service Systems", *Futures*, Vol. 32, No. 3-4, pp. 289-299.

Saunders, M. (1997), *Strategic Purchasing and Supply Chain Management* (2nd ed.), Financial Times Prentice Hall, London.

Schmidheiny, S. (1992), *Changing Course: A Global Business Perspective on Development and the Environment*, MIT Press, Cambridge, MA.

Seuring, S. and M. Muller (2007), "Core Issues in Sustainable Supply Chain Management: A Delphi Study", *Business Strategy and the Environment*, published online on 10 December.

Tukker, A. *et al.* (2006), *New Business for Old Europe*, Greenleaf Publishing, Cheltenham.

United Nations Environment Programme (UNEP) and United Nations Industrial Development Organization (UNIDO) (2004), *Guidance Manual: How to Establish and Operate Cleaner Production Centres,* UNIDO, Vienna.

UNIDO (2002), "The New Industrial Revolution: Michel Braungart at Venice II", *UNIDO Scope Weekly News*, 20-26 October, pp. 2-3, UNIDO, Vienna.

Ven, A.H. van de and T.J. Hargrave (2002), "Social, Technical and Institutional Change", in M.S. Poole and A.H. van de Ven (eds.), *Handbook of Organizational Change and Innovation*, Oxford University Press, Oxford.

Veleva, V. and M. Ellenbecker (2001), "Indicators of Sustainable Production: Framework and Methodology", *Journal of Cleaner Production*, Vol. 9, No. 6, pp. 519–549.

Weizsacker, E. von, A.B. Lovins and L.H. Lovins (1998), F*actor Four: Doubling Wealth, Halving Resource Use*, New Report to the Club of Rome, Earthscan, London.

Wired (2008), "BMW Hydrogen 7 Mono-Fuel Eats Smog for Breakfast", *Wired Magazine*, 16 April.

World Business Council for Sustainable Development (WBCSD) (1996), *Eco-efficient Leadership for Improved Economic and Environmental Performance*, WBCSD, Geneva.

World Commission on Environment and Development (WCED) (1987), *Our Common Future,* Oxford University Press, Oxford.

World Meteorological Organization (WMO) and United Nations Environment Programme (UNEP) (1998), *Scientific Assessment of Ozone Depletion: 1998*, WMO Ozone Report No. 44, WMO, Geneva.

WMO and UNEP (2006), *Scientific Assessment of Ozone Depletion: 2006*, WMO, Geneva.

World Bank (2007), *Global Economic Prospects 2007: Managing the Next Wave of Globalization*, The World Bank, Washington, DC.

Chapter 2

Applying Eco-innovation: Examples from Three Sectors

To better represent the contexts and processes that lead to eco-innovation, this chapter presents some illustrative examples of various eco-innovative solutions from three sectors: automotive and transport, iron and steel, and electronics. The primary focus of current eco-innovation efforts in these sectors tends to be technological advances in the form of product and process modifications or redesigns. However, some actors have started to explore more systemic eco-innovation through new business models and alternative modes of provision. Changes in organisational or institutional arrangements have acted as key drivers of technological development.

Introduction

As concerns over climate change and ecological depletion rise on corporate agendas, commercial interest in developing and applying innovative solutions across a broad range of areas is growing. As Chapter 1 argued, eco-innovation offers new perspectives on moving industrial production onto a sustainable path. However, companies do not appear to have a clear understanding of the concept and how it can be applied in their business operations. Many eco-innovations may have been realised unintentionally or without planning to reduce environmental impact. Consequently, and given the large variety and diversity of eco-innovations, the contexts and processes that lead to eco-innovation are not well known. This has hindered a more direct and focused approach to realising and promoting eco-innovation among industry, policy makers and researchers. Against this backdrop and to improve understanding of this issue, there seems to be a growing need to look closely at examples of eco-innovation.

This chapter presents illustrative, sector-specific examples of eco-innovation. The main aim is twofold. First, a number of examples of eco-innovation are taken from three sectors: the automotive and transport sector, the iron and steel sector, and the electronics sector.[1] Second, the development processes and characteristics of the eco-innovations from those industries are analysed using the typology of eco-innovation developed in Chapter 1 (Figure 2.1) which categorises eco-innovation in terms of target, mechanisms and impacts. The target refers to the area in which the eco-innovation takes place: products (goods or services), processes, marketing methods, organisational structures and institutional arrangements. The mechanism reflects the way in which the change in the eco-innovation target is brought about; it ranges from modifications and redesigns of the target to the use of alternative methods or techniques, to the creation of completely new elements. The impact reflects the environmental benefits to be achieved by the eco-innovation. The analysis aims at a better understanding of the diverse nature of the examples and their realisation.

Examples of eco-innovation from the three sectors are presented and analysed in the following sections on the basis of a general description, the process that led to their development, and an appraisal of their characteristics. Each of the three industry sections briefly summarises the sectoral eco-innovation characteristics. The chapter concludes with an overall synthesis of eco-innovation examples.

This chapter does not seek to give an exhaustive overview of eco-innovation in various sectors, nor are the examples meant to represent "best practices". Instead, its aim is to illustrate as far as possible the diversity of eco-innovation and of the contexts in which it occurs (Table 2.1). As such, it is an attempt to show how eco-innovation and its components fit into company's overall business activities.

Figure 2.1. The typology of eco-innovation

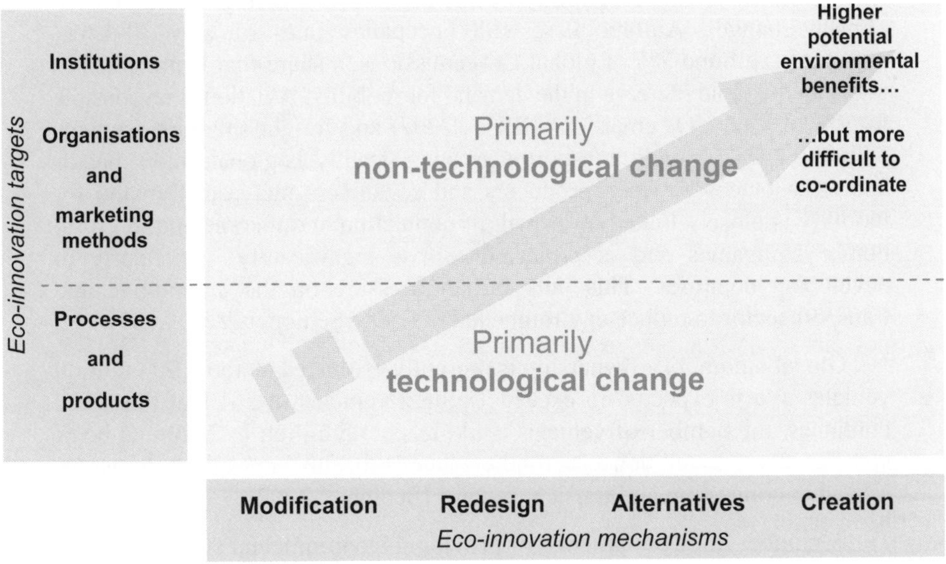

Eco-innovation targets

Institutions

Organisations and marketing methods

Processes and products

Primarily non-technological change

Primarily technological change

Higher potential environmental benefits...

...but more difficult to co-ordinate

Modification | Redesign | Alternatives | Creation

Eco-innovation mechanisms

Table 2.1. Examples of eco-innovation in three industry sectors

Sector and organisation	Eco-innovation example
Automotive and transport	
The BMW Group	Improving energy efficiency of automobiles
Toyota	Sustainable plants
Michelin	Energy-saving tyres
City of Paris & JC Decaux	Self-service bicycle sharing system
Iron and steel	
Siemens VAI, etc.	Alternative iron-making processes
ULSAB-AVC	Advanced high-strength steel for automobiles
Electronics	
IBM	Energy efficiency in data centres
Yokogawa Electric	Energy-saving controller for air conditioning water pumps
Sharp	Enhancing recycling of electronic appliances
Xerox	Print management services

Eco-innovation in the automotive and transport sector

Background

Overall, the transport sector accounts for 20% of global carbon dioxide (CO_2) emissions and is currently one of the fastest-growing contributors to climate change.[2] Automobiles, with occupancy rates of only 30-40%,[3] account for around 7% of global CO_2 emissions, a share that is projected to rise with the rapid increase in the demand for mobility. Aviation is responsible for 2% of total CO_2 emissions (IPCC, 2007) and freight shipping accounts for some 4.5%.[4] Freight transport is growing steadily as globalisation spreads and as distances between producers and consumers increase. Demand for mobility is closely linked to growth in population and income and in cross-border companies and is projected to rise significantly, particularly in developing countries. This puts further pressure on the automotive and transport sector to reduce environmental impacts (Schipper *et al.*, 2007).

Global automobile ownership is presently estimated at about 900 million vehicles and is expected to exceed 1 billion vehicles in 2010. If this trend continues, the number of vehicles could reach 1.5 billion in 2020 and be of significantly greater concern to the planet's health as well as to issues related to congestion and traffic accidents (Schipper, 2007).

A number of developments for providing environmental solutions in the automotive and transport sector have emerged over the past decade. For automobiles, this has led to significantly lower fuel requirements for a given horsepower and weight. But many of these gains have been offset by increasing energy demands in more powerful, larger and heavier new vehicles, particularly in the United States where the number of on-road sport utility vehicles (SUVs) has increased significantly (Schipper, 2007).

Improving the energy efficiency of automobiles – the BMW Group

The BMW Group, a German car manufacturer, has been engaged in eco-innovation to conserve resources and improve energy efficiency in automobiles, thus improving fuel economy for consumers while reducing the amount of CO_2 emissions from combustion. For example, its high-precision injection systems have enabled its four- and six-cylinder petrol engines to achieve fuel consumption levels during "lean operation" which could previously only be attained by diesel engines. BMW vehicles sold in Europe have been equipped with this system since 2007.

The BMW Group has also improved the fuel economy in its vehicles through better energy management. For example, the "auto start-stop" function switches the engine off automatically when the vehicle comes to a halt. Its brake energy regeneration technology makes use of both the braking and acceleration phases to charge the vehicle's battery and also works to reduce drag on the engine. Thus, as soon as the driver stops accelerating, kinetic energy is automatically harnessed and fed into the battery. By contrast, the alternator is disengaged during acceleration. This results in lower fuel consumption and maximum thrust when accelerating. Electric steering assistance and efficient, demand-controlled operation of fuel, coolant and oil pumps ensure that aggregates are only activated for as long as necessary. Active aerodynamics measures enable air flaps at the front of the vehicle to be opened for only as long as the engine requires air from outside for cooling purposes. This helps to speed up the warm-up phase and improves aerodynamics at the same time. In addition, the company developed a gear-shift indicator that provides the driver with real-time feedback on the optimum moment to change gears to conserve energy, in effect encouraging drivers to drive in a more fuel-efficient manner.

The company also now offers a comprehensive technology package that reduces both fuel consumption and exhaust emissions as a standard feature across all model series and car segments. By developing and implementing these solutions, the company has reduced CO_2 emissions from its own fleet in Europe by almost 27% between 1995 and 2008. In 2009, it announced that the first two BMW models with ActiveHybrid technology will be introduced to the market. Compared to other models powered solely by a conventional combustion engine, these hybrid models will reduce fuel consumption by up to 20%. For the long term, the company considers hydrogen as the preferred solution for sustainable mobility but is also exploring alternatives such as electric drive.

As part of their declared sustainability activities, the BMW Group is also working on the management of traffic and transport to improve the energy yield of all vehicles. Efforts in this area include improved management of traffic and parking (*i.e.* through better planning to secure free traffic flows while minimising the probability of congestion) and training programmes for fuel-save driving.

The process behind these product developments stems from the company's Efficient Dynamics Strategy, introduced in 2000 partly in response to the Kyoto Protocol. To support the strategy, the company established a separate division under its Development Office, which works in an integrated manner on issues related to vehicle energy management, aerodynamics, light-weight construction, performance and CO_2 emissions.

A key component of this strategy and of the company's approach to eco-innovation is the adoption of a life cycle perspective in product design. The company seeks to design cars with a view to conserve resources in relation to the production and use of the car, to secure the safety of drivers and passengers, and to increase recycling rates when the product reaches the end of its life. The strategy is also concerned with constant optimisation of all of its models, not just niche models as is often the case.

The continuous improvement in corporate environmental performance through this strategy has been facilitated by the use and ongoing updating of various indicators such as CO_2 emissions and fuel consumption throughout the model range. This ensures that the company has access to up-to-date performance assessments and thus promotes the idea of a more systematic and consistent approach to eco-innovation which is in line with eco-efficiency. Like other car manufacturers, the BMW Group is also engaged in improving the general efficiency of all production sites, with the help of an information system that collects data on some 150 environmental performance indicators.

Most eco-innovations implemented by the BMW Group involve technological advances across a range of product and process elements. Their realisation is nonetheless the outcome of consciously designed corporate strategies and various changes to the company's organisational structure. The company has used creative organisational procedures and processes to foster continuous improvements in various target areas of their products, guided by a life cycle perspective and the implementation of an information collection system. As noted, the eco-innovative process is underpinned by a separate division that works specifically to optimise product performance in a number of key environmental areas. The company also demonstrates eco-innovative efforts in the institutional sphere through its collaboration with governments and other stakeholders, including its training programmes for fuel-save driving and its work on establishing recovery centres to increase take-back and recycling rates.

Sustainable plants – Toyota

In moving towards sustainable manufacturing, the Japanese car manufacturer Toyota adopted the concept of "sustainable plants" with a view to creating production sites in harmony with their natural surroundings. The concept has given rise to a range of eco-innovative activities across three main areas. First, the company has sought to reduce its energy consumption by developing and implementing low-carbon production technologies and by daily *kaizen* (continuous improvement) activities. Second, it is increasing use of energy that stems from renewable sources. Third, it is actively involved with the local communities surrounding its production facilities on such issues as nature preservation. These local engagement activities have also been used to raise the environmental awareness of the company's employees.

In connection with the use of renewable energy sources, the company installed one of the largest photovoltaic power generation systems for automobile production at its Tsutsumi plant in Toyota City, Japan, where its Prius hybrid vehicles are manufactured. The system is comprised of 12 000 solar panels and covers an area equivalent to 60 tennis courts. With a rated output of approximately 2 000 kilowatts, the system can supply about half the electricity needed in the plant's assembly process.

The sustainable plant initiative at the Tsutsumi site also covers the conservation and rejuvenation of the surrounding eco-system. In this connection, the company has organised tree-planting events with trees native to the area, with the participation of local residents, employees and their family members. The company also plans to make use of technologies stemming from its biotechnology and afforestation businesses, for example by covering the walls and roofs of their automotive manufacturing plants with vegetation that can help to absorb emissions of nitrogen oxides (NOx) or the use on plant exteriors of photo-catalytic paint that can break down airborne NOx and sulphur oxides (SOx). In addition, the company intends to use the stream that runs through the plant grounds as a public gathering place for local people to appreciate and learn about the surrounding nature.

The tree-planting events are part of a broader company strategy to increase environmental awareness and understanding among employees as this can build a foundation for future improvements based on suggestions from employees. To facilitate such developments, the company introduced an "eco-point system" which gives employees points for providing ideas that help to reduce energy consumption or conserve the environment, or for partaking in environmental activities such as tree-planting events. Employees with outstanding performance receive awards.

Toyota's fourth Environmental Action Plan states its environmental responsibilities and the yearly targets to be achieved between 2006 and 2010. The action plan, which was first introduced in 1993, seeks to achieve a balance between the company's growth and harmony with society through specific actions, measures and goals in the areas of development and design, procurement, logistics and marketing, with a specific emphasis on four themes: energy and global warming, recycling of resources, substances of concern, and atmospheric quality. These themes help to guide the company in its efforts to develop technologies for sustainable mobility.

As part of the Environmental Action Plan, the company has employed life cycle assessment (LCA) techniques to identify the activities associated with its manufacturing of automobiles that consume the most energy. As the painting process was identified as a top contributor (together with machining and casting), efforts have been directed to developing a new painting technology.

On a more general scale, the company is experimenting with a number of different visions and sustainability initiatives using the Tsutsumi plant as a "prototype". Building on its experiences, the company designated four additional production sites located in the United Kingdom, France, Thailand and the United States to serve as prototypes for taking sustainable plant initiatives further. The plant in Thailand, which became operational in 2007, started recycling wastewater and has sent no waste to landfill from the start of its operations.

Toyota is known for its development and commercialisation of hybrid propulsion technology. But, it is also working to improve its production processes, including the substitution of conventional energy with renewable sources. Much of Toyota's eco-innovation can be described as redesign of products and processes and the use of alternative resources and methods.

One of the driving forces behind Toyota's eco-innovative developments is the company's strategy to achieve higher environmental performance. Its Environmental Action Plans have paved the way for greater focus on environmental and social aspects of automobile manufacturing and have been a driver for the "sustainable plants" concept which is now being implemented at selected plants.

Although these developments still are in a relatively early phase, they signify that Toyota is expanding its understanding of sustainability in a way that goes beyond the company's core business. This could help to create the circumstances for developing eco-innovative solutions that otherwise may not be considered. Covering the roofs of manufacturing plants with vegetation that absorbs NOx emissions and use of photo-catalytic paint with similar properties on the walls can serve as examples in this regard. Likewise, the company has engaged in eco-innovative institutional arrangements such as active involvement of local communities in order to preserve the ecosystems surrounding its factories.

Energy-saving tyres – Michelin

The supply of tyres to the automotive and transport industry involves several stages, including procurement of raw materials, such as natural rubber from hevea trees, and manufacturing and distribution of the tyres. While these processes are associated with a number of environmental challenges, the largest negative environmental impact is incurred during the use of tyres. It stems from "rolling resistance", which is linked to the demands of tyre performances such as grip and handling and contributes to fuel consumption and CO_2 emissions.

The first generation of Energy Saving products from Michelin, a French tyre manufacturer, was introduced in 1992 and the fourth generation of "green tyres" was launched in 2008. According to the company, the substitution of silica for carbon black in the latest generation has led to reductions in rolling resistance of nearly 20%. This translates into a reduction in fuel consumption of nearly 0.2 litres per 100 km in combined city and motorway driving (Michelin, 2008).[5] The company has also managed to prolong the mileage durability of their tyres significantly while maintaining or improving braking performance. Today, the company estimates that the fitting and use of their Energy Saving products have helped to save more than 10 billion litres of fuel and the emission of more than 26 million tonnes of CO_2.[6] At the same time, the material mass of production has also been reduced. Over the coming years, Michelin plans to work further on its next-generation tyre which promises further reductions in both rolling resistance and material mass. The company is currently engaged in establishing an industry standard for displaying information on rolling resistance of tyres. Such a standard does not currently exist.

The company used an extensive LCA of tyre production to learn that 86% of CO_2 emissions stem from the rolling phase, *i.e.* when the tyre is in use, and the remaining 14% from raw material production, tyre manufacturing, retail and disposal. It also learned that tyre rolling resistance accounts for as much as 20% of the fuel consumption of standard cars. For trucks, this proportion can reach more than 30%. The company estimated that 4% of all anthropogenic CO_2 emissions could be ascribed to rolling resistance of tyres. Considering its market share, Michelin tyres could account for 0.8% of total CO_2 emissions linked to human activity.

This LCA estimation essentially initiated the company's eco-innovation process and led it to look into how rolling resistance could be reduced to obtain higher fuel efficiency and thus lower the cost of mobility, while also causing less exhaust. The company found these objectives could be achieved by partly replacing carbon black, which is used as reinforcement filler in tyres, with silica.

According to the company, the development of its Energy Saving tyres was not an easy process because the substitution of silica for carbon black was a time-consuming and risky process which the company could not undertake alone. One of Michelin's suppliers of raw materials, however, was ready to undertake the task and worked with the company for a couple of years. Another factor was the priority given to the project by the company's top management and an investment in R&D of almost EUR 400 million.

Efforts to further increase tyre efficiency, in particular for trucks, are partly driven by the company's business model in the market for truck tyres. The company's Fleet Solutions programme is applying the product-service system (PSS) model (see Chapter 1) by selling "tyre maintenance services", calculated in kilometres driven, which include fitting, checking pressure, changing and replacing, and re-grooving and retreading tyres, etc. This creates incentives for the company to reduce the costs associated with retaining tyre ownership, compared with the conventional business model of selling the tyres.

The company's LCA of its manufacturing process not only gave Michelin a clearer picture of its direct and indirect environmental impacts and responsibilities but also enabled it to better target its R&D efforts to find cost-effective solutions that could improve environmental performance. However, owing to the relatively high-risk and time-consuming process of developing options for eco-innovation, the company modified its approach to conducting R&D by engaging in a collaborative research partnership with one of the company's raw material suppliers.

The company's engagement in establishing an industry standard for the display of information on rolling resistance of tyres signifies two additional eco-innovative elements: *i)* the company looks towards adopting a greener profile in its marketing strategy and positioning, and *ii)* it seeks to instigate a formal eco-innovative institutional change in the market that could help to increase consumer awareness of the relation between rolling resistance and fuel consumption, and eventually help to drive a change in purchasing behaviour.

The self-service bicycle sharing system in Paris – Vélib'

The relatively low air quality in Paris stems from its dense population and traffic. Despite a number of improvements across a wide range of pollutants, the Paris area still does not meet some national and European standards.[7] In further attempts to reduce traffic congestion and improve air quality, as well as to make the city a greener, quieter and more relaxed place, the City of Paris introduced a self-service bike-sharing system called Vélib' (for *vélo libre* – free bicycle) in the summer of 2007. The bicycle service builds on the success of a similar system introduced in Lyon in 2005.

The Vélib' system consists of some 1 750 stations located in conjunction with metro and bus stations and open 24 hours a day year round, each containing 20 or more bike spaces (Figure 2.2). This amounts to about one station every 300 metres throughout the inner city, with a total of 23 900 bicycles and 40 000 bicycle racks. Each station is equipped with an automatic rental terminal at which people can hire a bicycle through different subscription options. Subscriptions can be purchased for a small fee by the day, week or year and can be linked to the "swipe and enter" Navigo card used for the city's metro and bus system. By October 2009, the number of annual subscribers had reached 147 000, and between 65 000 and 150 000 Vélib' trips were being made each day. The system was extended to 30 towns in neighbouring suburbs by mid-2009.[8]

Figure 2.2. A self-service station of the Vélib' bicycle-sharing system in Paris

A subscription allows the user to pick up a bike from any station in the city and use it freely for 30 minutes. After that time a charge is incurred for additional time in chunks of 30 minutes. The payment scheme was designed to keep bicycles in constant circulation and increase sharing intensity. To facilitate circulation, bicycles are also redistributed every night to stations at which they are in particularly high demand. Real-time data on bicycle availability at every station is provided through the Internet and is also accessible via mobile phones.

The start-up financing for the Vélib' project, as well as full-time operation for ten years and associated costs, was entirely borne by the JC Decaux advertising company. In return, the City of Paris transferred full control of a substantial portion of the city's advertising billboards to this company. With this source of income, JC Decaux would expect to run a considerable profit in the third year of the Vélib' project, even though all income generated by the bicycle-sharing system itself goes to the City of Paris.

Overall, the system has been a great success and using Vélib' bicycles has also become fashionable. Part of this success is due to the system's design and application, with its strong focus on flexibility, availability and ease of use. The bicycles are built to be heavy and robust to increase their reliability and to minimise the risk of theft and vandalism.[9]

Building on the success of Vélib', the city is now planning to expand the programme to about 4 000 self-service electric hire cars. This new system, called Autolib', is expected to work on the same principles as Vélib', *i.e.* with drivers signing up for an annual subscription, including some form of free usage per vehicle or per day. Users will be able to reserve a car over the Internet 24 hours a day. The system is expected to be in place by the beginning of 2011.

The Vélib' system in Paris does not only aim at the supply of a flexible means of transport to reduce congestion. It is also a major part of the city's attempt to foster a more wide-ranging change in the population's general view of transport and in-city commuting. The target of the eco-innovative Vélib' system therefore takes an institutional or cultural focus, and the primary mechanism of change is the provision of an alternative means of transport.

The provision of alternative transport and its capacity to foster a cultural change nonetheless builds on a number of critical initiatives undertaken by the City of Paris. These include the careful planning of the many bicycle stations constructed throughout the city, the construction of dedicated bike lanes, as well as the restructuring of a number of roads to create a more bicycle-friendly environment. Moreover, processes such as nightly redistributions of bicycles to areas of high demand help ensure the system's functionality and provide for its flexibility. This is critical for inducing the intended cultural shift and a change in transport behaviour.

Overview of automotive and transport initiatives

The automotive and transport sector has taken several steps to reduce CO_2 emissions as well as other environmental impacts, notably those associated with fossil fuel combustion. Combined with growing demands for mobility, particularly in emerging economies, the eco-innovation initiatives have generally focused on increasing the overall energy efficiency of automobiles and transport, while increasing automobile safety. For the most part, eco-innovation in this sector has been realised through technological advances, typically in the form of modification and redesign of products or processes such as more efficient fuel injection technologies, better power management systems, energy-saving tyres, and optimisation of painting processes.

Yet, there are also indications that the understanding of eco-innovation in the automotive and transport sector is broadening and becoming increasingly integrated. Alternative business models and modes of transport such as the bicycle-sharing scheme in Paris are being explored by new players, as are brand new ways of dealing with pollutants from the manufacturing processes of automobiles. Several companies have taken initiatives to engage in both informal and formal institutional arrangements as a means to expand their environmental responsibilities.

Eco-innovation in the iron and steel sector

Background

Production of iron and steel is one of the most energy-consuming industrial activities, and issues relating to CO_2 emissions and energy efficiency are therefore of primary concern for the industry. By itself, the iron and steel sector accounts for some 7% of anthropogenic CO_2 emissions. This figure would increase to about 10% if mining and transport of iron ore are included (OECD, 2007). The industry is also significant for other environmental concerns such as waste treatment and the use of natural resources.

Steel is made from iron using two principal methods: the blast furnace/ basic oxygen furnace (BF/BOF) process, which accounts for two-thirds of world production and uses iron ore as the principal iron-bearing feedstock, and the electric arc furnace (EAF), which relies mostly on steel scrap.[10] Coke is a critical input in the BF/BOF steelmaking process; it is produced from hard coal and is needed to extract metallic iron from iron ore. Making coke poses significant environmental concerns at virtually every step of the production process and is, along with iron-making, often seen as the steel industry's greatest environmental challenge. Steelmaking via the EAF process is less polluting than the BF/BOF method but depends on the availability of scrap steel and consumes vast amounts of electricity, which creates environmental issues as well.[11]

The iron and steel industry has worked to improve environmental performance in recent years. The development of new production techniques has eliminated many energy-intensive steps in the steelmaking process and reduced emissions of air pollutants. Efforts to utilise waste heat and increase automation in production processes have raised fuel efficiencies and steel yields. Also, the development of new products such as high-strength and corrosion-resistant steels, and the increased recycling of by-products, have reduced the industry's environmental impact. Nevertheless, although CO_2

emissions per tonne of steel produced have declined noticeably, rapid growth in demand and in steel production has led to a 19% rise in total CO_2 emissions since 1990 (OECD, 2006). Growing demand for infrastructure, housing and rapid industrialisation in emerging economies, particularly China, has contributed to this development. Collaborative research initiatives are therefore actively searching for breakthrough technologies that could radically reduce CO_2 emissions.

Alternative iron-making processes

Over the last decades, R&D in the iron and steel industry has led to growing use of alternative iron-making methods known as direct smelting reduction processes. A number of such processes exist such as Corex, FASTEEL, FASTMET and HIsmelt. Corex is currently the most industrially and commercially advanced.

These processes differ from traditional iron-making by allowing for direct smelting of the iron using non-coking coal. They produce hot metal of equivalent quality to that produced in a conventional blast furnace. The Corex process does not completely eliminate the need for coke, but it reduces it significantly, thus lowering overall raw material costs and some of the negative environmental impacts associated with the coke-making process (Chatterjee, 2005).

Smelting reduction was initially conceived in Scandinavian countries, and the first attempt at a sustained process was made in 1938-39 in Denmark. Although interest in the process waned early on owing to technological advances in direct reduction technology, the technology was revived because of low productivity, product handling problems and high cost of production in the direct reduction process (Chatterjee, 2005).

The Corex process was developed by Austria's Voest-Alpine Industries (VAI) in the late 1970s. The technology was brought to the feasibility stage in the 1980s (Kastner, 2007) and the first Corex operating plant began production at Iscor (South Africa) in 1989. Four Corex plants were subsequently put into operation by Posco (Korea), Mittal Steel (South Africa) and Jindal (India). The technology, which is still being refined, has so far produced more than 25 million tonnes of liquid hot metal (Kastner, 2007). Although this is a relatively small amount compared with world's total pig iron production of 876 million tonnes in 2006, some observers expect the capacity of Corex plants to increase rapidly in the medium to long term. In November 2007, the world's largest Corex plant went into operation with a 1.5 million tonne operating capacity for China's Baosteel.

According to Siemens Metals and Mining, which merged with VAI in 2005 (now Siemens VAI), cost savings in hot metal production can be up to 20%, depending on the local site conditions. Also, emissions from Corex plants, which contain small amounts of NOx, sulphur dioxide (SO_2), dust, phenols, sulphides and ammonium, are below future expected European standards. By reducing the need for coking plants, the Corex process also reduces CO_2 emissions, potentially by 30%, according to Siemens (Wegener, 2007). Generation of wastewater from the Corex process is likewise lower than in conventional BFs. Waste plastics can also be fed directly into the Corex and other smelting reduction processes to reduce the fuel rate. Therefore, the expensive injection equipment used for this technique in conventional iron-making processes is not needed (Gupta, 2004).

Since the early 1990s, Siemens VAI and Posco's Research Institute of Industrial Sciences have been working on improving the Corex process. Finex, developed by Posco, completely eliminates the need for coke. This has the advantage of eliminating the intensive capital requirements associated with coking plants. Finex also allows the use of non-agglomerated iron ore fines in the iron-making process; this eliminates sintering and the need for a sinter plant. A recent demonstration showed that, compared to the BF, Finex reduces SOx by 92%, NOx by 96%, and dust emissions by 79% (IISI, 2006).

A further issue is the increasing scarcity of the hard metallurgical coal which is used as raw material for producing coke for the conventional iron-making route. This has led to an increase in the overall cost of raw materials for conventional BF iron-making, which amounts to 50-60% of total costs (Chatterjee, 2005). Highly capital-intensive coke plants and the negative environmental impacts associated with coke-making have stepped up economic and environmental pressure on the iron and steel industry to develop alternative iron-making routes. In addition, the smelting reduction process makes it possible to meet the increased demand for a cost-efficient capacity to produce smaller quantities of hot metal (Chatterjee, 2005).

The above-mentioned factors have been among the primary drivers behind the further development of the Corex and Finex technologies. To this end, Siemens VAI is actively engaged in co-operative partnerships with several universities and research centres on the development of the Finex process.

The eco-innovation characteristics of the direct smelting reduction processes can generally be described as a process modification in one of the steel-making routes. However, when compared to iron-making in the conventional BF, direct smelting reduction is more progressive as it replaces coke with coal to extract iron. Not only does this eliminate environmental impacts associated with coke making, it also makes steel production more flexible. From the perspective of iron making, this eco-innovation is better described as a process redesign that enables steel making at smaller scales and on the basis of alternative raw materials.

Innovative developments have nevertheless also occurred in the conventional blast furnace route, including improved process technology, and better design and engineering of the equipment involved. These changes have continued to raise the competitiveness of the traditional iron-making process and have made it harder to develop and commercialise new smelting reduction processes. One of the most important factors for the future development of Corex and Finex is therefore the ability to compete in terms of cost.

Advanced high-strength steels for automobiles

Steel is an important raw material for the automotive sector, and 13-14% of the world's steel is used to manufacture motor vehicles. In Germany and the United States, the automotive industry accounts for more than 20% of steel consumption. In China, it is as low as 3%, partly owing to the massive quantities of steel used in construction.

From an environmental perspective, the heavier the automobile, the more energy required for propulsion and thus higher emissions. The iron and steel industry, together with the automotive industry, has therefore been developing advanced high-strength steel in order to manufacture lighter cars that increase fuel efficiency and lower exhaust emissions. It is estimated, for instance, that for every 10% reduction in vehicle weight, the fuel economy (measured by litres of fuel per 100 km driving distance) is improved by between 1.9% and 8.2% (worldsteel, 2008), depending on adjustments made to the vehicle's power train.

The total weight of a typical five-passenger family car is 1 260 kg, of which 360 kg for the car body when using conventional steel. If other parts that use steel are included, 55% of a typical car's weight is due to steel, according to the World Steel Association. By using advanced high-strength steel (at little additional cost compared to conventional steel), the overall weight saving could reach nearly 120 kg, or 9% the vehicle's total weight. If the weight is reduced, the power train can also be downsized without any loss in performance, thus leading to additional fuel savings. Moreover, with high- and ultra-high-strength steel components, such vehicles rank high on crash safety and require less steel for construction.

Box 2.1. Achieving fuel efficiency through innovative structural design using advanced high-strength steel

The Loremo (low resistance mobile) is the work of a German entrepreneurial company focused on the innovative use of advanced yet standard materials and engine technologies to create light-weight vehicles with low air resistance, without compromising passenger protection.

The Loremo's body is constructed using advanced high-strength steel with a linear cell structure. This means that the steel structure is uninterrupted on both sides over the entire length of the car, including a part in the middle. The car therefore does not have any doors and passengers enter the car by raising the hood, which includes the windscreen and steering column – see picture). The steel structure is zinc-plated to prevent corrosion and the car does not require painting as no steel is visible from the outside. Where strength is not needed, construction is based on thermo-plastic materials. The car is also designed for easy recycling.

The structural design ensures high safety standards and also makes it significantly lighter than conventional cars. With a weight of around 550 kg, a two-cylinder turbo diesel engine and a highly aerodynamic design, the smallest of the models can reach a maximum speed of 160 km/h and travel 100 km on about two litres of diesel.

This eco-innovation has been achieved by rethinking how cars are conceived and constructed. Originally, this car was intended as an affordable means of transport in emerging markets, but Loremo also plans to sell it to the European market in light of rising concerns over global warming. Mass production of the car is planned for 2011. The company is also working on the development of electric and hybrid versions. Loremo expects to sell the smallest model for less than EUR 15 000.

Source: Loremo website *www.loremo.com* and communication with the authors.

Other technologies have been developed to produce lighter and stronger auto bodies. Hydro-form tubing is a process for shaping hollow tubes, under high pressure, into light and strong one-piece shapes that can replace standard auto parts (see Box 2.1). Another technology, laser-welded automotive blanks, can make entire side panels in one operation by pre-welding sheets with high-speed lasers prior to forming. This allows for an optimal distribution of steel components in the car-making process, *i.e.* it uses the strongest steel where it is most needed and lighter steel elsewhere (IISI, 2002).

The above developments began with the introduction of new legislative requirements on motor vehicle emissions in 1993 in the United States. These intensified the pressure on industry to reduce the environmental impact associated with the use of automobiles. In response, industry formed the Ultra-Light Steel Auto Body (ULSAB) initiative, an international collaborative venture by vehicle designers and a number of steelmakers from around the world to develop stronger and lighter auto bodies. This venture led to the ULSAB Advanced Vehicles Concept (ULSAB-AVC) which aimed to showcase the latest high-technology steel grades for automotive applications. The Future Steel Vehicle (FSV) is the latest in the series of auto steel research projects. It combines global steel makers with a major automotive engineering partner and aims to demonstrate safe, light-weight steel bodies for future vehicles that reduce GHG emissions over the life cycle of the vehicle.

In 1999, the ULSAB-AVC carried out a proof-of-concept experiment for the application of advanced high-strength steel to automobiles, thus providing automakers with a way to reduce emissions while also producing safe, efficient and affordable cars. Demonstrations of other technologies to produce stronger and lighter auto bodies followed. Continuing efforts by the iron and steel industry to conduct R&D in these areas also stem from the industry's attempt to strengthen steel's competitive advantage over alternative materials such as aluminium. In 2005, the ULSAB-AVC received the Alliance to Save Energy's 2005 Stars of Efficiency Award in recognition of its advances in developing solutions for vehicle energy efficiency (AISI, 2005).

The target of these eco-innovative efforts was a steel product that would allow for the manufacture of a strong but light automobile body. This led to the development of advanced high-strength steel which can be described as a modification or a redesign of existing components and production methods.

However, the development of high-strength steel took form through the establishment of the ULSAB and later the ULSAB-AVC, a showcase and research consortium of vehicle designers and steelmakers. Active involvement in this cross-sectoral arrangement allowed the iron and steel industry to improve its knowledge and understanding of how steel is viewed by one of its major customers, and facilitated the realisation of mutual and environmental benefits through active collaboration.

The establishment of the ULSAB and the ULSAB-AVC can be classified as a formal institutional eco-innovation. It may point the way to similar arrangements for achieving other gains in the future. Already the ULSAB has evolved into a number of sister bodies occupied with research and demonstration of advanced high-strength steel in the production of other automobile components such as closures and suspensions.

Box 2.2. Ultra-low carbon steelmaking

The Ultra-Low Carbon Dioxide Steelmaking (ULCOS) programme was launched in 2004 as a co-operative R&D consortium of 48 companies and organisations from 15 European countries. Its aim is to reduce CO_2 emissions from steel production by at least 50% compared to today's best methods.

Its research activities started with a feasibility study of more than 80 technologies. Among these, four promising breakthrough technologies were identified for further R&D on the basis of significant CO_2 reduction and an examination of various process routes, depending on where and when they would be used. The research has also identified a number of almost mature technologies that can deliver small reductions in CO_2 emissions. These are now being developed outside the ULCOS programme.

Source: ULCOS website, *www.ulcos.org.*

Overview of iron and steel initiatives

The iron and steel industry has made significant progress in recent years to increase its environmental performance through a number of energy-saving modifications and redesigns of various production processes. These efforts have been driven by strong pressure on the industry to reduce pollution and by the increasing prices and scarcity of raw materials. Most eco-innovative initiatives in the iron and steel sector have therefore focused on technological product and process advances.

However, as for the automotive and transport sector, the engagement of the iron and steel industry in various institutional arrangements laid the foundation for many of these developments. The development of advanced high-strength steel, for example, was made possible through an international collaborative arrangement between vehicle designers and steelmakers and enabled the production of stronger steel for the manufacturing of lighter and more energy-efficient automobiles. Another very important factor is the changing economic situation of the industry, notably the increased cost and availability of raw materials such as coke.

Eco-innovation in the electronics sector

Background

While the automotive and transport sector and the iron and steel sector are widely regarded as major sources of CO_2 emissions, the electronics sector is also responsible for a large share of global energy consumption. The increasing consumption of electronic products also constitutes an increasing problem in terms of waste. This is due not only to growing demand for various consumer electronics and appliances, but also to the increasing incorporation of electronics in other goods.

At the same time, however, electronics also has significant potential for helping to reduce environmental impacts from different activities and industries. In the United States, for example, it is estimated that the power consumption of corporate computer servers and data centres could be lowered from current levels by an estimated 56% by 2011 by adopting energy-efficient methods and technologies (EPA, 2007).

In general, a number of eco-innovative designs and physical products have been developed to make more efficient use of energy, reduce CO_2 emissions and deal more effectively with equipment waste (e-waste). In response to consumer demand, electronics manufacturers have sought to create products with reduced energy consumption while simultaneously increasing their products' marketing and functional value (EIU, 2007). In recent years, the issue of recycling has also received increasing attention.

Energy efficiency in data centres – IBM

Maintaining central facilities that contain critical components essential to the running of many organisations requires substantial amounts of energy. Data centres consume a considerable amount of energy and can be up to 40 times more energy-intensive than conventional office buildings. In the United States, for example, the demand for power and cooling processes by data centres is estimated to have more than doubled between 2001 and 2006, and in 2006 data centres represented about 1.5% of the country's entire consumption of electricity, the equivalent of the energy consumed by about 5.8 million average households in the country (EPA, 2007; WBCSD, 2008).

The energy efficiency of a data centre is typically referred to as the data centre infrastructure efficiency (DCIE) and is measured as the energy consumption of the information technology (IT) equipment relative to the facility's total energy consumption. Good performance in this metric is a

DCIE of more than 60%. However, a study by IBM, an American IT company, revealed that the average DCIE was only 44%. In 2007, IBM and associated business partners therefore announced a project under which the company would invest USD 1 billion to deliver technologies, products and services to radically improve the energy efficiency of its clients' and its own operations, products and services. Named "Project Big Green", it includes a five-step approach to sharply decrease the energy consumption of data centres and thus lower both CO_2 emissions associated with energy usage and clients' energy costs. The five steps are described as:

1. *diagnose*: evaluate energy consumption of existing facilities: energy assessment, virtual 3-D power management and thermal analytics;
2. *build*: plan, build or update to an energy-efficient data centre;
3. *virtualise*: virtualise IT infrastructures and deployment of energy-saving special purpose processors;
4. *manage*: seize control with power management software;
5. *cool*: exploit liquid cooling solutions inside and outside the data centre.

IBM manages more than 740 000 square metres of data centres around the world and energy use has become a significant factor in operational costs and the ability to increase capacity and capability. A central feature of Project Big Green is the consolidation of 3 900 distributed servers on 33 System z servers in IBM data centres around the world. This is expected to save as much as 119 000 megawatt-hours (MWh) a year, enough electricity to power about 9 000 average US homes for a year. Through improvements in data centre energy efficiency, the company expects to double its IT capacity in data centres over the next three years without increasing energy use. Improving the energy efficiency of data centres starts with an assessment of existing data centre energy use.

To optimise energy usage in existing data centres, the company developed Mobile Monitoring Technology (MMT) to analyse the thermal profile of an operating data centre, identify "hotspots" and provide recommendations on improving the thermal profile. This technology has been offered as an "energy management service" to clients who wish to reduce their energy costs. The data collected by the MMT is processed in a specialised modelling tool to develop a three-dimensional rendition of the thermal and flow characteristics of the data centre. The results of the model are used to calculate six energy efficiency metrics: horizontal and vertical hotspots, non-targeted air flow, temperature variations in computer room air conditioning (CRAC) units or in plenum discharges, and flow blockage. The metrics point to opportunities to improve energy use in the data centre. Each metric has a corresponding set of easily implemented improvements which can be used to improve performance, often with little or no investment. Best practice assessments, which provide a similar but less data-driven analysis, also help optimise data centre energy use.

IBM is currently expanding its focus with a view to applying a green approach throughout the organisation. This is leading to more diverse offerings for clients, ranging from consulting for sustainability strategies and a greener supply chain to a holistic portfolio of "software for a greener world", which also focuses on reducing people's workload through productivity enhancements.

Influenced by concern, particularly in larger companies, over energy issues such as growing energy costs, capital requirements for building new data centres, poor power management and lack of electricity in general, the company's corporate strategy has turned towards leadership in energy management for the industry and its clients. These developments served as the original foundation for Project Big Green.

Project Big Green, along with a number of specific initiatives such as its research and consulting services for water management, also benefited tremendously from the company's Internet discussions, dubbed "innovation jams". In 2006, clients, employees and their families were brought together in two worldwide collaborative brainstorming sessions or jams. More than 150 000 people from 104 countries suggested more than 46 000 ideas. In a second phase, ten business opportunities, one of which is the Big Green Innovation initiative, were approved for further development (Davies, 2007). The initial work of Big Green Innovation, which has focused on data centres, accelerated the deployment of the MMT and an energy management business service.

To gain market acceptance of the MMT, the company engaged in a number of new marketing initiatives. Two of these were seen as essential to the success of the project and the establishment of the company's energy management service. The first was the demonstration of savings by active operating data centres. Here, the company teamed up with PG&E, a holding company of energy firms, and IBM Integrated Technology Delivery, the company's business unit which supplies data centre services, which were willing to participate in the testing. The second initiative was the demonstration of the technology's advantageous pay-back scheme and its ability to free up sufficient capacity to support further business.

IBM's development of the MMT can be classified as the creation of a new technology application as it provides a new tool for identifying and assessing how energy consumption in existing data centres can be reduced through modifications and redesigns or through the implementation of alternative equipment. While this eco-innovative technological development is a leap forward in itself, it is also a fundamental building block of another of the company's eco-innovative initiatives, the adoption of an alternative business model based on the provision of energy management services.

Indeed, the company's approach to environmental stewardship has moved over the years from sharing its experiences with external organisations and clients to combining green efforts and cost-cutting initiatives with business opportunities and new sources for revenue. Project Big Green and the company's recent energy management services, software portfolio for "going green", and its expanding consulting capabilities to help clients with energy and environmental issues across their operations illustrate this trend.

In its work leading to Project Big Green and the development of the MMT, the use of innovation jams also illustrates IBM's active engagement in the creation of novel eco-innovative institutional arrangements. Eco-innovative efforts in the form of alternative marketing strategies were also undertaken by testing and demonstrating the capabilities and cost effectiveness of the MMT through collaboration with clients.

Energy-saving controller for air conditioning water pumps – Yokogawa Electric

It has been said that for every dollar spent on powering a server, another dollar is spent on cooling it (Mehta, 2006). In view of the quantities of electricity consumed by servers mentioned above, cooling is another valuable target for energy saving. The same is true for air conditioning, which consumes vast amounts of energy to maintain regulated temperatures.

Air conditioners function by driving hot or cold water through piping structures to units located on each level of the building. The amount of cold water varies according to the desired temperature relative to the outside temperature. However, despite variances in the amount of water required, conventional air conditioners maintain operation at the pressure required to meet maximum heating and cooling demands. Consequently, vast amounts of energy are wasted. For example, research from Japan's Building Energy Managers' Association found that half of the office building energy in Japan is spent on air conditioning (Yoshida, 2006). The growing prevalence of air-conditioned buildings means that the total amount of energy spent on providing air conditioning is rapidly raising global CO_2 emissions.

To help reduce energy consumption from air conditioning, Yokogawa Electric, a Japanese manufacturer, developed a new technology called Econo-Pilot which controls the pumping pressure of air conditioning systems in a sophisticated manner. This innovation easily enables large energy savings as it can be applied to existing air conditioning systems, so that there is no need to buy new cooling equipment. The technology has been widely used in equipment factories, hospitals, hotels, supermarkets and office buildings (see Figure 2.3).

Figure 2.3. Econo-Pilot energy-saving control system for air conditioning water pumps

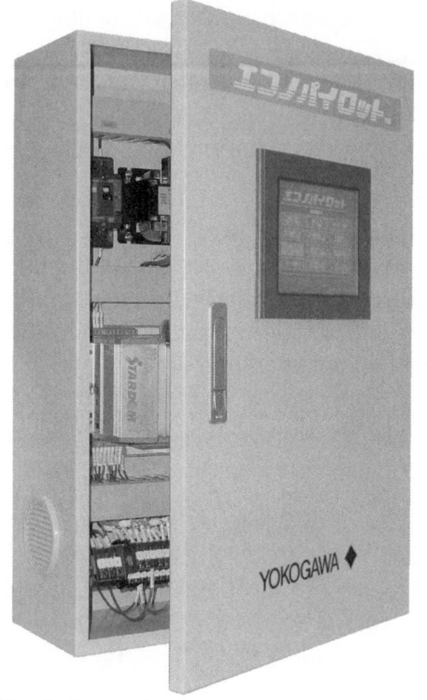

Source: Yokogawa Electric Corporation.

Based on changes in the flow rate, the controller calculates the minimum pressure necessary by means of data processing capabilities equivalent to those of a personal computer. Econo-Pilot's pressure control diminishes the redundant energy consumed when high pressure is continually maintained and significantly reduces the pump's electricity consumption. In many cases, it can reduce annual pump power consumption by up to 90%. The actual reduction in percentage terms varies depending on factors such as the type of air conditioning system in place and the type of pump control system used before Econo-Pilot was installed.

Yokogawa's eco-innovation grew out of a desire to fulfil public commitments to tackling global warming and to meet ISO 14001 environmental management system certification by reaching yearly targets for improvement. A prolonged recession in Japan made saving on energy a high priority for customers but building owners were not financially able to undertake large-scale renewal of equipment. Under these circumstances, the company saw an opportunity in the need for a way to significantly reduce costs without great expense.

Based on research revealing that air conditioning consumes half of a building's total energy consumption, the company sought to create a simple, inexpensive and low-risk control mechanism that could eliminate wasteful use of energy. The resulting product was the Econo-Pilot, which could be installed easily and inexpensively. The purchaser could expect a significant reduction in electricity consumption.

Econo-Pilot was based on technology devised jointly by Yokogawa with Asahi Industries Co. and First Energy Service Company. It was developed and demonstrated through a joint research project with the New Energy and Industrial Technology Development Organization (NEDO), a public organisation established by the Japanese government to co-ordinate R&D activities of industry, academia and the government. The NEDO undertakes research to develop new energy and energy-conservation technologies and works on their validation and implementation. After the demonstration and piloting of the technology, various functions were incorporated to complete the final product.

Yokogawa's development of Econo-Pilot represents the creation of an eco-innovation based on technological advances. From a broader perspective, however, it is best classified as a modification of conventional air conditioning systems. This is underlined by the fact that the Econo-Pilot is applied to existing air conditioning units and does not constitute an alternative or a new way of cooling.

Yokogawa's eco-innovative developments have also taken other forms, as illustrated by the company's organisational commitments to ISO 14001 certification, which have led it to pursue various environmental improvements in a more targeted manner. The company has also been engaged in collaborative research with other companies to develop its eco-innovative technology. In addition, its participation in an institutional collaborative arrangement, which included a public research organisation, was pivotal in the demonstration and pilot phase of the technology.

Enhancing recycling of electronic appliances – Sharp

In recent years, liquid crystal displays (LCDs) have replaced conventional cathode ray tubes (CRTs) in a variety of application areas. As an increasing number of LCDs are coming to the end-of-life phase, waste management of LCDs is a growing environmental concern. The major methods available to deal with redundant LCDs had been incineration or landfill, both of which cause safety and environmental hazards. Incineration of LCDs emits volatile products and residues. Research has shown that the backlight of many old LCDs contains mercury, which has damaging accumulative effects on the human body and the environment. Therefore, landfill is also ecologically damaging.

In the United States, of the 2.25 million tonnes of TVs, cell phones and computer products ready for end-of-life management, 18% were collected for recycling and 82% were disposed of primarily in landfills.[12] According to a study from Stanford Resources (San Jose, California), more than 2.5 billion LCD units were disposed of in 2003, with the annual increase estimated at 15%. As a result, the need to develop technologies to reduce the environmental impact is becoming urgent.

Since 2002, Sharp, a Japanese manufacturer, has been working on a corporate-wide project to develop a recycling technology for LCD TVs and other LCD applications. The company also set guidelines for the safe removal of mercury backlights in LCD TVs and LCD panels. In 2007, the company disassembled LCD TVs of all sizes to identify problems in the disassembly process. Using the knowledge gained through this activity, proof-of-concept experiments for recycling were implemented in 2008. The operation of a safe, efficient flat-panel TV disassembly and recycling line started in April 2009.

The company has also been working on technologies for recovering and recycling plastics. In 1999, it started developing the technology for closed-loop material recycling, an original technology for re-using plastics recovered from TVs, air conditioners, refrigerators and washing machines in new consumer electronics for the Japanese market. This technology was implemented in 2001 and the company has increased its use of recycled plastic every year. In 2008, the use of recycled plastics reached about 1 050 tonnes, up 100% from 2005.

Together with Aqua Tech Co., Sharp also developed a proprietary technique for recovering and recycling indium, a rare metal, contained in the transparent electrodes in LCD panels. This simple process uses common chemicals and eliminates the need for large energy expenditures. The company has completed proof-of-concept tests using large-scale prototype equipment and will move towards actual recovery operations.

While developing a recycling technology, the company has also engaged in a co-operative project involving five companies[13] to collect and recycle used electrical appliances. They formed a consortium to facilitate collection and retrieval of four types of appliances designated under the Japan's Home Appliance Recycling Law (TVs, air conditioners, refrigerators and washing machines).[14] They now operate 190 designated sites for picking up old appliances and 18 sites for recycling. In 2005, approximately 1.3 million

Sharp home appliance units were recovered and recycled through this system. Similarly, together with other personal computer (PC) manufacturers, Sharp formed a partnership with Japan Post Service Co. for the collection of discarded PCs at more than 20 000 post offices around Japan. Parallel to its efforts in Japan, Sharp USA together with Panasonic and Toshiba launched a nation-wide recycling programme in the United States. From January 2009, consumers have access to 280 recycling sites throughout the United States with hundreds more planned for the next three years.

Sharp's eco-innovative efforts to enhance the recycling rates of various electronic appliances partly grew out of the company's strategy and commitment to achieve a high level of environmental awareness in all corporate activities. This "Eco-Positive Strategy" covers four areas: technologies, products, operations and relationships.

As part of the "products" area of its environmental strategy, the company is working to develop recycling processes for products at their end-of-service life based on three goals: *i)* to improve the recycling rate and aim for zero landfill disposal; *ii)* to improve the efficiency of the recycling system to reduce recycling costs; and *iii)* to incorporate recycling technologies into the development and design of products. These objectives have directed much of the company's R&D effort, as well as its collaborative activities, in developing recycling programmes and electronic components and products.

It is important to note that the company's work on recycling has also been strongly affected by regulatory reforms. The Home Appliance Recycling Law, which came into force in 2001, has provided major Japanese electronics companies with the impetus to build recycling plants and to construct the necessary infrastructure for effective recycling operations.

Sharp's development of technologies that enhance or enable the recycling of various materials and components can essentially be classified as process modifications. At the same time, these process modifications have laid the foundation for the company to become more fully engaged in eco-innovative recycling efforts by means of other eco-innovative targets and mechanisms. This is exemplified by its work on constructing an infrastructure for enhanced recycling.

The sector-wide partnership with Japan Post Service Co. for collecting end-of-life computer equipment illustrates the company's engagement in eco-innovative institutional arrangements. The same can be said of its collaboration with various other electronics companies, both in Japan and in the United States, to establish a broader recycling system for electrical appliances.

Print management services – Xerox

Despite the growth in digital communications and promises of paperless offices, printed documents continue to occupy a major part of day-to-day business operations in many companies. For many years, print volumes have continued to grow, as has the inefficiency associated with printing jobs. To meet company demands for better cost control and management of increasingly complex printing environments, Xerox, a US printer manufacturer, introduced Managed Print Services (MPS). In essence, MPS implies a shift in the company's traditional business model from selling printing devices to supplying document services by providing customers with tailored solutions for document assets and infrastructure management.

The focus of the MPS business model is an enterprise-wide print management service to cut costs by minimising the energy consumption of printing devices, providing optimal maintenance of the equipment and reducing associated capital requirements by retaining ownership. The company also seeks to centralise the printing administration to facilitate continuous improvements in the printing infrastructure and to help client companies control their printing costs more effectively through the introduction of a pay-per-use scheme which makes it possible to keep track of all printing expenses. These services have been offered through four key phases: assessment, optimisation, implementation, and maintenance upgrades.

The first step of MPS involves calculating the "hard costs" (such as printing devices and copying equipment) and "soft costs" (such as energy and ink usage, support and maintenance) of printing by evaluating documents and workflows in the office in order to learn the total cost of the client's printing activities and equipment ownership. Xerox then works to ascertain key areas for improvement, to detect overworked and neglected devices, and pinpoint opportunities to eliminate excess expenditure. Then, an optimisation plan is designed to create the most efficient workspace layout to economise energy and maximise efficiency. Solutions can involve upgrading old equipment to more energy-efficient devices or reducing and redistributing current devices for better user and usage placement. Xerox takes responsibility for the step-by-step implementation of the upgrading plan to centralise management and device monitoring. In addition, it retains responsibility for maintaining all equipment and software, replenishing supplies, and ongoing monitoring to ensure maximum benefits and viability of the printing infrastructure.

The company also developed software programmes that allow better monitoring of networked printers and multi-function products. These send e-mail reports stating how many documents each device created, the ink or toner levels, and the due dates for scheduled maintenance. To better track and raise awareness of the environmental impact of printing, the company made a "sustainability calculator" that allows clients to measure the waste and CO_2 emissions associated with powering printers, copiers, fax machines and multi-function devices. It also enables them to compare the environmental impact of printing single-sided and double-sided documents and that of different types of ink.

In association with its MPS, Xerox has developed a number of technologies which help cut costs while reducing the environmental impact of printing and copying. For example, it developed the Solid Ink Colour Technology, which eliminates cartridges in laser printers and thus generates 90% less waste while cutting costs and improving reliability. Parallel advances have been made in toner technology with a new emulsion aggregation agent that facilitates toner particle grinding and uses 40-45% less toner mass per page while reducing overall energy consumption by 15-22% per pound of toner manufacturing. One of more novel developments, though still unavailable on the market, is an erasable paper on which printed images are erased after 16-24 hours and can be used again.

Xerox's development and offering of its specialised MPS partly arose from the increasing pressure for companies to cut costs associated with their IT infrastructure. In most cases such efforts have focused on improvements related to networks, servers, storage capacities, software solutions and computers, and less attention has been paid to the improvement of printing and copying infrastructures. Because printing and copying activities in most companies are spread out across business units and locations, they constitute fragmented and independent "islands" in the IT infrastructure. This creates a significant market opportunity. Indeed, most companies are unable to state accurately how many printers they own, how many pages are printed every day, and how much their printing activities cost the company as a whole. Consequently, many companies have under-utilised devices or dated equipment which is costly to run. The costs can be considerable when including power consumption, maintenance, change of printing heads, support, etc.

Many of Xerox's other eco-innovative developments, such as the solid ink technology, stem, at least partly, from the company's alternative business model which essentially has internalised a number of costs associated with printing and copying. In short, the lower the costs of installation, operation, maintenance, support and replacement, the higher the potential profit for the company. This translates into strong incentives for the company to minimise waste streams, material usage and energy consumption and to design products for easier remanufacturing and recycling.

Xerox's development and supply of MPS is an example of eco-innovative business models, as it derives environmental benefits by internalising the costs of using, maintaining, and refitting printing and copying machines with the manufacture of the equipment itself. In comparison with the conventional business model of selling physical printers and copying machines, Xerox focuses its eco-innovation mechanism on providing an alternative product through its document management services.

Overview of electronics initiatives

The electronics sector has so far directed most of its eco-innovation efforts towards the reduction of energy consumption. However, as consumption of electronic equipment continues to grow, companies are also seeking more efficient ways to deal with the waste generated.

Like the automotive and transport sector and the iron and steel sector, most eco-innovations in the electronics sector have focused on technological advances in the form of product and process modifications or redesigns. Similarly, developments in these areas have been built upon a number of eco-innovative organisational and institutional arrangements. Some of these arrangements have, perhaps unsurprisingly, been among the most innovative and forward-looking in terms of how eco-innovation may be approached in the future. A notable example is the use of large-scale Internet discussion groups by IBM, with the capacity to harness innovative ideas and knowledge among thousands of people.

Alternative business models, such as the provision of product-service solutions rather than physical products, have also been increasingly applied in the sector. This has been exemplified by new services in the form of improving the management of energy usage in data centres as well as printing and copying infrastructures.

Conclusions

This chapter presents illustrative examples of various eco-innovative solutions from the automotive and transport, iron and steel, and electronics sectors, in an attempt to show how eco-innovation fits within overall business activities. The examples were based on the typology of eco-innovation developed in Chapter 1 (see also Figure 2.1).

In the automotive and transport sector, eco-innovative solutions have generally focused on reducing CO_2 and other emissions associated with fuel combustion, driven by growing concern over climate change and increasing demand for mobility in developing countries. Eco-innovation has therefore targeted technological advances in products and processes, typically through their modification and redesign. Eco-innovative arrangements of an organisational or institutional character, based on both alternative and new means, have nevertheless accompanied many of the technological developments, and have started to be addressed more explicitly in the industry's approach to sustainability.

The iron and steel sector has faced issues relating to the availability of raw materials and concerns about its environmental impact. Against this backdrop, eco-innovation targets have mostly been determined by product and process optimisation, and mechanisms have typically been characterised by modification and redesign. Yet, the industry has also been actively engaged in eco-innovative institutional arrangements, such as with the automotive industry as well as with various research institutes.

The electronics sector has started focusing its eco-innovative efforts on the use phase of its products, typically by achieving lower energy consumption through product modification and redesign. These activities have mostly been driven by the industry's high market and consumer exposure, combined with growing concern over environmental impacts. The rising consumption of electronic products has also shifted the industry's focus to product and process design, as well as their engagement in institutional arrangements, for example, to enhance product recycling possibilities. Some players in the sector have also adopted product-service systems as alternative business models and some are applying creative collaborative institutional arrangements such as large-scale Internet brainstorming events.

Figure 2.4. Mapping primary focuses of eco-innovation examples

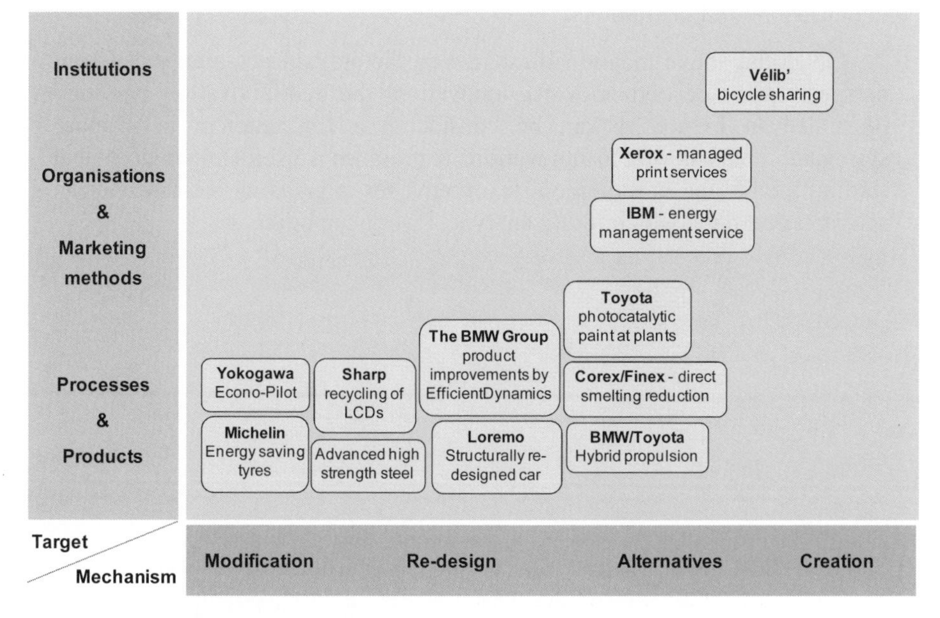

Note: This map only indicates primary targets and mechanisms that facilitated the listed eco-innovation examples. Each example also involved other innovation processes with different targets and mechanisms.

While this chapter's review shows that understanding eco-innovation processes and characteristics is complex, Figure 2.4 nevertheless attempts to map the examples covered in this chapter according to the eco-innovation typology.

Generally, it can be said that the primary focus of current eco-innovation in these sectors tends to be technological developments and advances, typically with products or processes as the target of the eco-innovation, and with modification or redesign as the eco-innovation's principal mechanism. Nevertheless, even with a strong focus on technological advances, it is also clear that a number of changes have served as drivers of the development of eco-innovation. In many of the examples examined in this chapter, these changes have been either organisational or institutional. They include the establishment of separate environmental divisions to monitor and improve overall environmental performance and help direct R&D efforts, and the establishment of inter-sectoral or multi-stakeholder collaborative research networks.

Since companies tend to deal with eco-innovative ideas and activities in very diverse ways, eco-innovative solutions take many different forms. Hence, the heart of an eco-innovation cannot necessarily be adequately represented by a single set of target and mechanism characteristics. Instead, eco-innovation seems best examined in terms of an array of characteristics ranging from modifications to creations, across products, processes, organisations and institutions.

Given the above-mentioned interacting factors and potentially different perspectives associated with eco-innovation, the eco-innovation typology illustrated in Figure 2.1 can be considered a first attempt at a more systematic analysis of eco-innovation. It provides a useful methodological starting point and a common taxonomy for appraising eco-innovation activities and upon which future analytical work can build.

Notes

1. The examples were chosen as a result of the initial sectoral focus of this project and of previous work by the OECD's Structural Policy Division under the auspices of the OECD Steel Committee. The information comes primarily from a company-level questionnaire survey conducted by the OECD between July and September 2008 in co-operation with the Business and Industry Advisory Committee to the OECD (BIAC) and the Advisory Expert Group for this project. This information is supplemented with input from the focus group meetings of corporate experts from the electronics and automotive and transport sectors organised by the OECD during the International Conference on Sustainable Manufacturing held in September 2008 in Rochester, New York, and at the DG Enterprise and Industry of the European Commission in Brussels in November 2008. Other information is drawn from publicly available sources including company websites and corporate sustainability reports. For iron and steel sector, the information mainly draws on the OECD report on environmental challenges in the industry prepared for the OECD Steel Committee (OECD, 2007).

2. World Resources Institute (WRI), Climate Analysis Indicators Tool (CAIT), *http://cait.wri.org*.

3. Air Transport Action Group (ATAG), *www.atag.org*.

4. International Air Transport Association (IATA), *www.iata.org*.

5. Based on an ISO test conducted by Germany's TÜV SÜD Automotive in 2007 on store-bought 175/65 R14 and 205/55 R16 tyres produced by five major manufacturers.

6. Estimates are from *www.compteur-vert-michelin.com*.

7. Common Information to European Air (CITEAIR), *www.airqualitynow.eu*.

8. Personal communication with the City of Paris and JC Decaux.

9. Even though Paris has put great effort into making the city more bicycle-friendly, such as by constructing more than 400 kilometres of bicycle lanes (600 km by the end of 2013), the city lacks a well-established and well-behaved cycling culture as the rapid growth in bicycle usage has led to more accidents. Furthermore, JC Decaux has complained to the City of Paris about the high level of vandalism and thus high maintenance costs. The city authority records that 16 000 bicycles have been vandalised and 8 000 have disappeared since the system's inception in 2007 (City of Paris, 2009). To address these issues, the city authority initiated a new traffic safety campaign at the end of 2008 and an anti-vandalism campaign in May 2009.

10. The heavy-polluting open hearth furnace (OHF) is still in use and accounts for about 2% of world production but has become obsolete in most countries.

11. For a more complete description of the environmental challenges facing the iron and steel industry, see OECD (2007).

12. US Environmental Protection Agency (EPA), "Statistics on the Management of Used and End-of-Life Electronics", *www.epa.gov/waste/conserve/materials/ecycling/manage.htm*.

13. The five companies include Fujitsu General, Hitachi Appliances, Mitsubishi Electric, Sanyo Electric and Sony.

14. This law requires manufacturers and importers to recycle used air conditioning units, televisions, refrigerators and washing machines. It also requires retailers to retrieve and send them to original manufacturers or importers for recycling. Consumers are required to pay fees to finance these activities before or at the time of disposing a used appliance.

References

American Iron and Steel Institute (AISI) (2005), "ULSAB-Advanced Vehicle Concepts (ULSAB-AVC) Recognized with Energy Efficiency Award", press release, 13 October, AISI, Detroit, MI.

Chatterjee, A. (2005), "A Critical Appraisal of the Present Status of Smelting Reduction", *Steel Times International*, Vol. 29, No. 5, pp. 36-42.

City of Paris (2009), "Un Vélib', ça se protégé!", 28 May, City of Paris, France, *www.paris.fr*.

Davies, J. (2007), *Big Green: IBM and the ROI of Environmental Leadership*, AMR Research Report, AMR Research, Inc., Boston, MA.

Economist Intelligence Unit (EIU) (2007), *IT and the Environment: A New Item on the CIO's Agenda?*, EIU, London, *www-05.ibm.com/no/ibm/environment/pdf/grennit_oktober2007.pdf*.

Environmental Protection Agency, United States (EPA) (2007), *Report to Congress on Server and Data Center Energy Efficiency*, EPA, Washington, DC.

Gupta, S.K. (2004), *Corex Process: One of the Dynamic Routes For Gel Making with Special Reference to the Success of JVSL*, Indian Steel Joint Plant Committee, Kolkata, *http://jpcindiansteel.nic.in/corex.asp*.

International Iron and Steel Institute (IISI) (2002), *Industry as a Partner for Sustainable Development*, IISI, Brussels.

IISI (2006), *Steel: The Foundation of a Sustainable Future Steel – Sustainability Report of the World Steel Industry 2005*, IISI, Brussels.

Intergovernmental Panel on Climate Change (IPCC) (2007), "Summary for Policymakers", in B. Metz *et al.* (eds.), *Climate Change 2007: Mitigation, Contribution of Working Group III to the Fourth Assessment Report of the Intergovernmental Panel on Climate Change*, Cambridge University Press, Cambridge, *www.ipcc.ch/pdf/assessment-report/ar4/wg3/ar4-wg3-spm.pdf*.

Kastner, W. (2007), "Next Generation Corex Technology", *Metals & Mining*, February, pp. 24-25, Siemens VAI, Linz.

Mehta, S.N. (2006), "Server Mania", *Fortune*, August 7, pp. 69-75.

Michelin (2008), *Michelin Energy Saver: Press Kit*, presented at the Geneva International Motor Show, March, Geneva.

OECD (2006), "Present Policy Approaches to Reduce CO_2 Emissions in the Iron and Steel Sector", internal working document for the Steel Committee.

OECD (2007), "Environmental Challenges in the Iron and Steel Industry", internal working document for the Steel Committee.

Schipper, L. (2007), *Automobile Fuel Economy and CO_2 Emissions in Industrialized Countries: Troubling Trends through 2005/6*, EMBARQ, World Resources Institute, Washington, DC, *http://pdf.wri.org/automobile-fuel-economy-co2-industrialized-countries.pdf.*

Schipper, L., M. Cordeiro and W. Ng (2007), "Measuring the Carbon Dioxide Impacts of Urban Transport Projects in Developing Countries", paper presented at Transportation Research Board 87th Annual Meeting, 13-17 January, Washington, DC.

Wegener, D. (2007), "Emission Reduction in Industry and Infrastructure Will Be Driven Mostly by Energy Savings", presentation at Siemens Media Summit, Siemens, Munich.

World Business Council for Sustainable Development (WBCSD) (2008), *IBM: Data Center Energy Efficiency, Case Study*, WBCSD, Geneva.

World Steel Association (worldsteel) (2008), *An Advanced High-Strength Steel Family Car*, Environmental Case Study: Automotive, WSA, Brussels, *www.worldsteel.org*.

Yoshida, Y. (2006), "Development of Air Conditioning Technologies to Reduce CO_2 Emissions in the Commercial Sector", *Carbon Balance Management*, Vol. 1, No. 12.

Chapter 3

Tracking Performance: Indicators for Sustainable Manufacturing

Measurement helps manufacturing companies to define objectives and monitor progress towards sustainable production. This chapter reviews the existing sets of indicators that help them track and benchmark their environmental performance. There is no ideal single set of indicators which covers all of the aspects which companies need to address to improve their production processes and products/services. An appropriate combination of elements of existing indicator sets could help them gain a more comprehensive picture of economic, environmental and social effects across the value chain and product life cycle.

Introduction

Sustainable production, which emerged as an aspect of sustainable development at the United Nations Conference on Environment and Development (UNCED) held in Rio de Janeiro in 1992,[1] has become a key component of business strategies in recent years. Many manufacturers have had to deal with increasing environmental regulatory measures but have also realised economic benefits by reducing resource use and waste, operating their production processes more efficiently, and promoting environmentally sound products and services.

To promote sustainable production and eco-innovation, companies and policy makers need data in order to understand issues relating to existing production systems, define specific objectives, and measure progress. Their desire for metrics is grounded on the proposition that in a business setting "what you don't measure, you can't manage". However, the measurement and monitoring of business activities is not necessarily easy at the practical level. This is partly because the concept of sustainable development is too multi-faceted for simple quantitative measurement and because an emphasis on environmental and social aspects often runs counter to the conventional government and industry agenda of economic growth.

This chapter reviews the existing sets of indicators that have been used to help track and benchmark different aspects of companies' performance in order to improve production processes and products/services towards sustainable development. It:

- introduces the sets of indicators for sustainable production that have typically been used by companies and business associations in the manufacturing sectors;
- analyses these indicator sets in terms of their effectiveness in realising and advancing sustainable production and eco-innovation based on the defined criteria;
- provides background information on what the OECD could contribute to improving indicators for sustainable production among OECD and non-OECD economies.

The following section explains why indicators are necessary for companies' operations and management decision making. The subsequent section categorises the sets of indicators; each category's characteristics are then presented, with appropriate examples, and analysed on the basis of certain predefined criteria. Current applications of indicators in manufacturing companies and companies' views on further development based on a questionnaire survey and focus group interviews are then described. Finally, a synthesis of sustainable manufacturing indicators is presented.

It should be noted that this chapter's scope is limited to the application of sustainable production indicators in manufacturing industries (*i.e.* sustainable manufacturing indicators), even though the categorisation and analysis of the existing indicator sets may also be applicable to other industries. It also emphasises environmental aspects of sustainable production.

How can indicators help sustainable manufacturing?

Functions of indicators

The management of complex issues in organisations requires ways to represent these issues with simple units of measurement so that timely decision making is possible. This condensed information for decision making is called indicators (Olsthoorn *et al.*, 2001). Body temperature is an example of an indicator we regularly use as it provides critical information on our physical condition. Likewise, indicators provide information about phenomena that are regarded as typical of and/or critical to the quality of target issues.

Companies use indicators to set targets and then monitor progress. Interpretation is easier if it is possible to set targets for the indicators themselves, as they help decision makers visualise the actions they will need to focus on in the future. Indicators can go beyond simple data and illustrate trends or cause-and-effect relationships between different phenomena. Typically, indicators have the following three key objectives:

- **To raise awareness and understanding.** Indicators are useful for describing baseline and current conditions (*e.g.* the amount or magnitude of something) and the performance of a system. They can provide the common language for describing a particular system that is needed for effective and clear communication among interested parties (McCool and Stankey, 2004).

- **To inform decision making.** Indicators help to make decisions and move analysis to a diagnostic mode, as they can be a source of real-time feedback on performance. They can reveal what additional analysis may be needed to better understand a phenomenon. For example, an observed change may be an aberration or derive from systemic change. In either case, further monitoring and research are needed to understand the underlying causes.

- **To measure progress towards established goals.** Indicators offer a measure of the effectiveness of actions in moving a system towards a more desirable state. For example, if body temperature decreases after taking a medicine, we conclude that the medicine has been effective in combating the disease. To accomplish this objective, indicators should provide an ability to assess cause-and-effect relations.

The emergence of sustainable manufacturing indicators

In the past decades, sustainable development indicators have been developed at the global, regional, country and local levels. They help policy makers and the public to understand the linkages and trade-offs between economic, environmental and social values in order to evaluate the long-term implications of current decisions and behaviour and to monitor progress towards sustainable development goals by establishing baseline conditions and trends.

While in the past the behaviour of companies with respect to sustainable development was mainly directed by government, some companies have begun to recognise the potential competitive advantages and other business benefits of adopting a more conscious and proactive approach to sustainable development. The understanding and management of environmental and social performance is a prerequisite for realising sustainable development and should therefore be a basic asset for a company's competitiveness. At the same time, in the wake of a series of corporate scandals such as oil spills and sweatshop labour, there has been significant pressure from the public for businesses to be more accountable and transparent in their activities. Shareholders are also becoming increasingly vocal in their demands for non-financial information on business activities. The idea that organisations should be held accountable for their economic, environmental and social impacts is often referred to as corporate social responsibility (CSR).

There is increasing need for methods to make objective measurements and benchmark companies' performance with respect to the environment and sustainable development. Once companies recognise the need to embrace sustainable development, they need to learn how to achieve it. The development of sustainability indicators related to products/services and production processes is a good way for companies to incorporate the goal of sustainability into management decision making (Schwarz *et al.*, 2002). Better understanding of the links between sustainability performance, competitiveness and business success could enable profit-oriented organisations to realise their "win-win-win" potential (Schaltegger and Wagner, 2006).

Existing sets of sustainable manufacturing indicators

Categories of indicators and review criteria

A number of manufacturing companies have already started to use certain sets of indicators to measure and monitor the state and progress of their operations (sites/facilities, products/services) as well as their management (company as a whole) towards realising and advancing sustainable production. These indicator sets have been developed by various organisations, including public authorities, industry associations and non-governmental organisations (NGOs), and many companies have also developed their own indicator sets according to their needs. As a result, there exist a multitude – and diversity – of indicator sets for sustainable manufacturing around the world.[2] However, they do not appear to have been comprehensively categorised.

The OECD (2005) provides a definition of environmental indicators by distinguishing between *parameter*, *indicator* and *index*.[3] In reality, however, most companies combine various parameters and indicators and apply them as a "set" in order to understand the state and progress of their sustainability performance. To analyse the use of different metrics by companies, this chapter covers all types of metrics applications, referred to as "sets of indicators" (or indicator sets). On the basis of a multitude of indicator sets drawn from publicly available information such as academic literature and corporate reports, the following categories were identified (Table 3.1):

- individual indicators,
- key performance indicators (KPIs),
- composite indices,
- material flow analysis (MFA),
- environmental accounting,
- eco-efficiency indicators,
- life cycle assessment (LCA),
- sustainability reporting indicators, and
- socially responsible investment (SRI) indices.

This categorisation focuses on ways for companies to organise data and measurements in order to understand the overall performance of their manufacturing processes and products/services. The above categories were selected as: *i)* focused on sustainable production in manufacturing industries;

ii) applied by many companies in practice; and *iii)* not strongly based on another set of indicators. This categorisation is mainly based on how companies call the different indicator sets and distinguish them from other sets of indicators with different characteristics.

Table 3.1. Indicator sets for sustainable manufacturing

Category	Description	Similar indicators or examples
Individual indicators	Measure single aspects individually	Core set of indicators Minimum set of indicators
Key performance indicators (KPIs)	A limited number of indicators for measuring key aspects that are defined according to organisational goals	
Composite indices	Synthesis of groups of individual indicators which is expressed by only a few indices	
Material flow analysis (MFA)	A quantitative measure of the flows of materials and energy through a production process	Material balance Input-output analysis Material flow accounting Ecological footprint Exergy; MIPS; Ecological rucksack
Environmental accounting	Calculate environment-related costs and benefits in a way similar to financial accounting system	Environmental management accounting Total cost assessment Cost-benefit analysis Material flow cost accounting
Eco-efficiency indicators	Ratio of environmental impacts to economic value created	Factor
Life cycle assessment (LCA)	Measure environmental impacts from all stages of production and consumption of a product/service	Carbon footprint Water footprint
Sustainability reporting indicators	A range of indicators for corporate non-financial performance to stakeholders	GRI Guidelines Carbon Disclosure Project
Socially responsible investment (SRI) indices	Indices set and used by the financial community to benchmark corporate sustainability performance	Dow Jones Sustainability Indexes FTSE4Good

The following section describes the major characteristics of each category of indicator sets, accompanied by examples of their application in boxes. Each category is analysed in detail to evaluate its effectiveness in initiating and advancing corporate sustainable manufacturing practices. Whereas each company's operating environment is unique, in order to ensure an objective analysis, benchmarking criteria that are generally desirable for companies' usage of indicator sets are identified:

- **Comparability for external benchmarking.** Companies are facing intense competition and need to perform better than their competitors and the industry average and improve their performance over time. In the absence of benchmarks, companies have little idea of how they compare with competitors (Matthews and Lave, 2003). This applies equally to environmental and other sustainability performance. In fact, the lack of common measurement for sustainable production has hampered the adoption and dissemination of sustainable manufacturing practices (OECD, 2006). Even though some companies have established their own benchmarks for continuous improvement, these tend to be tailored to each company and may not allow for comparison within and across sectors. A recent study demonstrates that comparability is the single most important characteristic of environmental performance indicators. There is a growing need among investors, communities and consumers for comparable standardised sustainability indicators that make it possible to compare companies and products/services (Veleva and Ellenbecker, 2001).

- **Applicability for SMEs.** For small and medium-sized enterprises (SMEs) sustainable manufacturing indicators should be easy to apply in terms of cost and labour for data collection as well as ease to understand and use. A large majority of manufacturers in the supply chain are SMEs, but they are generally much less likely to embark on environmental improvement programmes than larger companies. A survey of a cross-section of SMEs in Australia shows that SMEs tend to consider environmental issues as a potential cost and not as a market opportunity. They also tend to take environmental measures only in response to threats of penalties by authorities and usually respond with "end-of-pipe" pollution control solutions (Rao *et al.*, 2006). Many have not established indicator systems owing to a lack of resources such as finance, personnel, time and technical knowledge as well as motivation and awareness.

- **Usefulness for management decision making.** Sustainable manufacturing indicators need to be able to provide useful information for management decision making. This criterion implies that indicators should be simple to interpret and comprehend and useful for decision making because they reflect the objectives of the organisation and its mechanisms. In the same way, indicators can serve to evaluate results delivered by management.

- **Effectiveness for improvement at operational level.** Sustainable manufacturing indicators should also be able to provide information that reflects improvement at the operational level, *i.e.* production processes and manufacturing of products/services. This requires integrity of information regarding all important operations. The indicators can be a guide to improvement when they help to understand day-to-day operations. This criterion also implies clarity and timelyness in the implementation of possible improvements.

- **Possibility of data aggregation and standardisation.** This characteristic implies that the indicators can be stacked in a standardised form so that the information collected at the production process, site or corporate level can be used at broader levels – within a sector, a country or around the globe. Data collected in stackable units can be aggregated for comparison and evaluation of diverse aspects of businesses. Considering that supply chains span facility fences, company walls and national boundaries, stackable indicators would also be useful for evaluating effects throughout the value chain.

- **Effectiveness for finding innovative products/solutions.** In relation to eco-innovation, it would be ideal if indicator sets also enabled companies to identify more innovative products and solutions to various sustainability challenges. Comparing experimental results with accumulated data can reveal which products/solutions are more sustainable.

Analysis of indicator sets: their characteristics and effectiveness

Individual indicators – measuring single items

A set of individual indicators is a simple compilation of single indicators, which measure diverse aspects of sustainable development either quantitatively, with standard units, such as dollar/euro, gram/tonne and litre/cubic metre, or rates (percentage), or qualitatively, with descriptions. These indicators measure individual aspects of the system, such as amounts of water use, energy consumption, waste generation, and recycling rate. Each indicator is basically independent and benchmarked separately. This set of indicators is the most one commonly used by companies as the first step in developing and applying sustainability indicators for each facility and/or company. A set of individual indicators can also be applied to sectors, countries and the world.

Table 3.2. Examples of individual indicators

Operating performance indicator (OPI)	Management performance indicator (MPI)	Environmental condition indicator (ECI)
Raw material used per unit of product (kg/unit)	Environmental costs or budget ($/year)	Contaminant concentrations in ambient air (μg/m^3)
Energy used annually per unit of product (MJ/1 000 l product)	Percentage of environmental targets achieved (%)	Frequency of photochemical smog events (per year)
Energy conserved (MJ)	Number of employees trained (% trained/to be trained)	Contaminant concentration in ground- or surface water (mg/l)
Number of emergency events or unplanned shutdowns (per year)	Number of audit findings	Change in groundwater level (m)
Hours of preventive maintenance (hours/year)	Number of audit findings addressed	Number of coliform bacteria per liter of potable water
Average fuel consumption of vehicle fleet (l/100 km)	Time spent to correct audit findings (person-hours)	Contaminant concentration in surface soil (mg/kg)
Percentage of product content that can be recycled (%)	Number of environmental incidents (per year)	Area of contaminated land rehabilitated (hectares/year)
Hazardous waste generated per unit of product (kg/unit)	Time spent responding to environmental incidents (person-hours per year)	Concentration of a contaminant in the tissue of a specific local species (μg/kg)
Emissions of specific pollutants to air (tonnes CO_2/year)	Number of complaints from public or employees (per year)	Population of an specific animal species within a defined area (per m^2)
Noise measured at specific receptor (dB)	Number of fines or violation notices (per year)	Increase in algae blooms (%)
Wastewater discharged per unit of product (1 000 l/unit)	Number of suppliers contacted about environmental management (per year)	Number of hospital admissions for asthma during smog season (per year)
Hazardous waste eliminated by pollution prevention (kg/year)	Cost of pollution prevention projects ($/year)	Number of fish deaths in a specific watercourse (per year)
Number of days air emissions limits were exceeded (days/year)	Number of management-level staff with specific environmental responsibilities	Employee blood lead levels (μg/100 ml)

Source: Putnam and Keen (2002), "ISO 14031: Environmental Performance Evaluation", draft submitted to *Journal of the Confederation of Indian Industry*, Altech Environmental Consulting, Toronto.

Box 3.1. A core set of individual indicators

A standard, practical set of individual indicators for sustainable manufacturing can be useful. For example, the Lowell Center for Sustainable Production at the University of Massachusetts, Lowell, suggests 22 core single indicators as commonly applicable to manufacturing companies. These core indicators include not only environmental aspects but also some social aspects such as community and labour issues:

Aspect	Indicator	Metric	Level
1. Energy and material use	(1) Fresh wastes consumption	litres	Level 2
	(2) Materials used (total and per unit of product)	kg	Level 2
	(3) Energy use (total and per unit of poduct)	kWh	Level 2
	(4) Percent energy from renewables	%	Level 2
2. Natural environment (including human health)	(5) Kilograms of waste generated before recycling (emission, solid and liquid waste)	kg	Level 2
	(6) Global warming potential (GWP)	tons of CO_2 equivalent	Level 3
	(7) Acidification potential	tons of SO_2 equivalent	Level 3
	(8) kg of PBT chemicals used	kg	Level 3
3. Economic performance	(9) Costs associated with EHS compliance (*e.g.* fines, liabilities, worker compensation, waste treatment and disposal, remediation)	USD	Level 1
	(10) Rate of customer complaints and returns	number of complaints/ returns per sale	Level 2
	(11) Organisation's openness to stakeholder review and participation in decision-making process (scale 1-5)	number (1 to 5)	Level 2
4. Community development and social justice	(12) Community spending and charitable contributions as percent of revenues	USD	Level 2
	(13) Number of employees per unit of product or dollar sales	number/USD	Level 2
	(14) Number of community-company partnerships	number	Level 2
5. Workers	(15) Lost workday injury and illness case rate	rate	Level 2
	(16) Rate of employees' suggested improvements in quality, social and EHS performance	number of suggestions per employee	Level 2
	(17) Turnover rate or average length of service of employees	rate (years)	Level 2
	(18) Average number of hours of employee training per year	hours	Level 2
	(19) Percent of workers, who report complete job saticsfaction (based on questionnaire)	%	Level 3
6. Products	(20) Percent of products designed for disassembly, reuse or recycling	%	Level 4
	(21) Percent of biodegradable	%	Level 4
	(22) Percent of products with take-back policies in place	%	Level 4

…/…

Box 3.1. A core set of individual indicators *(continued)*

The Lowell Center also provides a hierarchy of five levels of indicators relative to the basic principles of sustainability to provide a tool for organisations to measure the effectiveness of their sustainability efforts. The lower levels of the hierarchy cover basic elements of sustainability. Level 1 covers compliance with regulations and industry standards, while Level 2 measures individual company efficiency and productivity. At Levels 3 and 4, companies have to look beyond their own organisational boundaries and consider the impacts of suppliers and distributors. This hierarchy emphasises that the development of indicators for sustainable production is not static but a continuous and evolutionary process of setting goals and performance measurement.

Source: Veleva and Ellenbecker (2001), "Indicators of Sustainable Production: Framework and Methodology", *Journal of Cleaner Production*, Vol. 9.

ISO 14031, an international standard for environmental performance evaluation, provides a standard process for measuring an organisation's environmental performance against its environmental policy, objectives, targets and other criteria, in line with the ISO 14001 environmental management system (EMS) standard. This standard also categorises different types of individual indicators. It distinguishes between environmental condition indicators and environmental performance indicators, and subdivides the latter into management performance indicators and operational performance indicators (Putnam and Keen, 2002). Table 3.2 gives examples of these indicators.

There is no limitation on the number of individual indicators to be used. This depends on what the relevant companies consider appropriate to obtain an overview of their performance with respect to sustainable development. However, since it can be resource-intensive to measure a large number of aspects and difficult to make a balanced and timely judgement, a small number of individual indicators may be selected as a core or minimum set of indicators (Box 3.1).

In terms of comparability, individual indicators are in principle unsuitable because they are applied to a large number of routine corporate procedures and can be created for each company according to its needs. If a sector could agree on a core set of indicators, this would greatly facilitate sector-level benchmarking among companies.

For SMEs, individual indicators are the most familiar and can be easily utilised for internal evaluation. They can be adopted without the organisational analysis and complicated calculations needed for key performance indicators or composite indices. However, since SMEs would have difficulty collecting data for many items, the number of indicators must be limited.

From the viewpoint of management decision making, individual indicators would not help management to understand the full picture since they present a wide variety of data independently. To be useful for management decisions, priority issues for management need to be identified and the number of individual indicators should be restricted. Individual indicators cannot identify links between environmental performance and financial outcomes, which management tends to need to make decisions on environmental investment.

For improvement at the operational level, individual indicators can only apply to a few selected environmental aspects. As indicators are monitored independently, the fact that improvement in one area may lead to deterioration in others can make this issue difficult to handle.

If consensus can be reached among concerned parties (*e.g.* members of a sector association) on the units of data and if organisational boundaries and a system to avoid double counting are properly established, individual indicators can be used for data aggregation and standardisation.

With regard to finding innovative products/solutions, individual indicators can be used only when companies focus on a few environmental attributes. As the focus on a single item might lead to overall deterioration of environmental performance, the use of individual indicators for product and process development is not advisable.

Key performance indicators – monitoring progress towards corporate goals

Key performance indicators (KPIs) are a set of quantitative and qualitative measurements defined by an organisation to measure progress towards its goals. KPIs are expressed by numbers or values which can be compared to an internal or external target for benchmarking and give an indication of the organisation's performance. These values can relate to data collected or calculated from any process or activity (Ahmad and Dhafr, 2002). What distinguishes KPIs from other indicator sets is their focus on organisational goals. If properly defined, KPIs can serve as a useful diagnostic tool to learn which measures are most effective. Any metrics can be selected to illustrate factors that are critical for assessing the success of the organisation. KPIs are in principle applicable to any organisation that seeks to improve its sustainability performance. They may differ depending on the organisation's structure and strategy.

KPIs usually involve long-term considerations and require an analysis of the organisation's mission and the identification of its stakeholders and organisational goals. KPIs can be helpful for managers who have to handle complex sustainability issues. A clear understanding of both the drivers of performance and the effects of that performance on various stakeholders may allow for better integration of the information into routine decision making and the institutionalisation of social concerns throughout the organisation (Epstein and Roy, 2001).

Figure 3.1. A model scheme for building KPIs

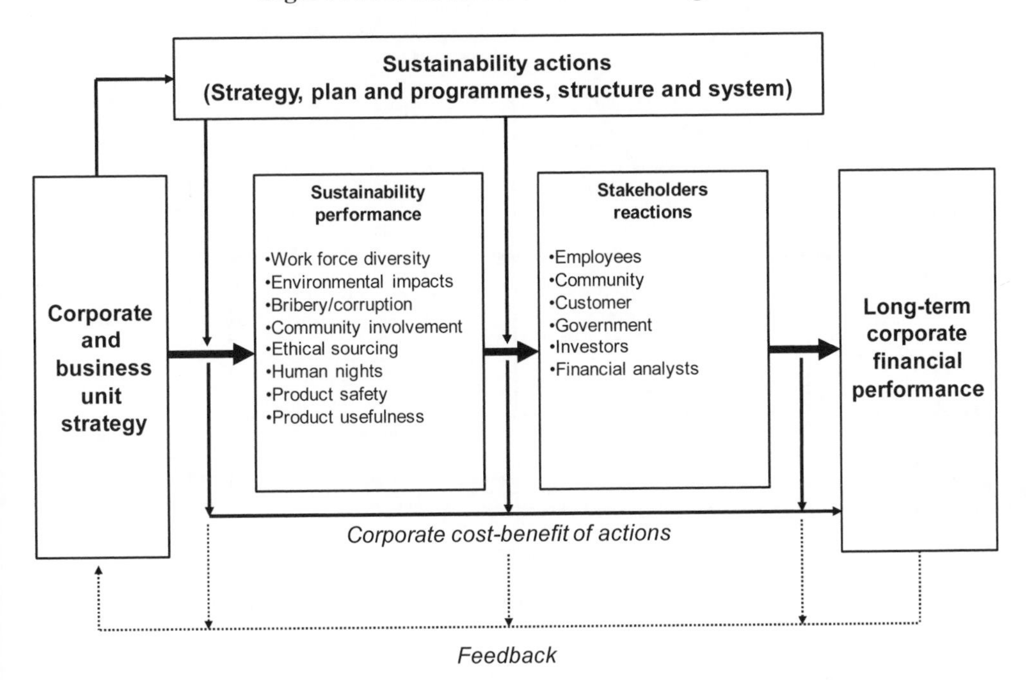

Source: Epstein and Roy (2001), "Sustainability in Action: Identifying and Measuring the Key Performance Drivers", *Long Range Planning*, Vol. 34.

Epstein and Roy (2001) present a model scheme for developing KPIs (Figure 3.1). It focuses on the relations between a company's strategy and actions for sustainable development, its sustainability performance, stakeholder reactions, and long-term financial performance (Box 3.2). The authors suggest establishing KPIs in each of these five areas so that the company can monitor whether and how its sustainability actions can improve sustainability as well as financial performance. Figure 3.2 shows an example of KPIs developed based on this model.

Figure 3.2. Example of a set of KPIs based on the model

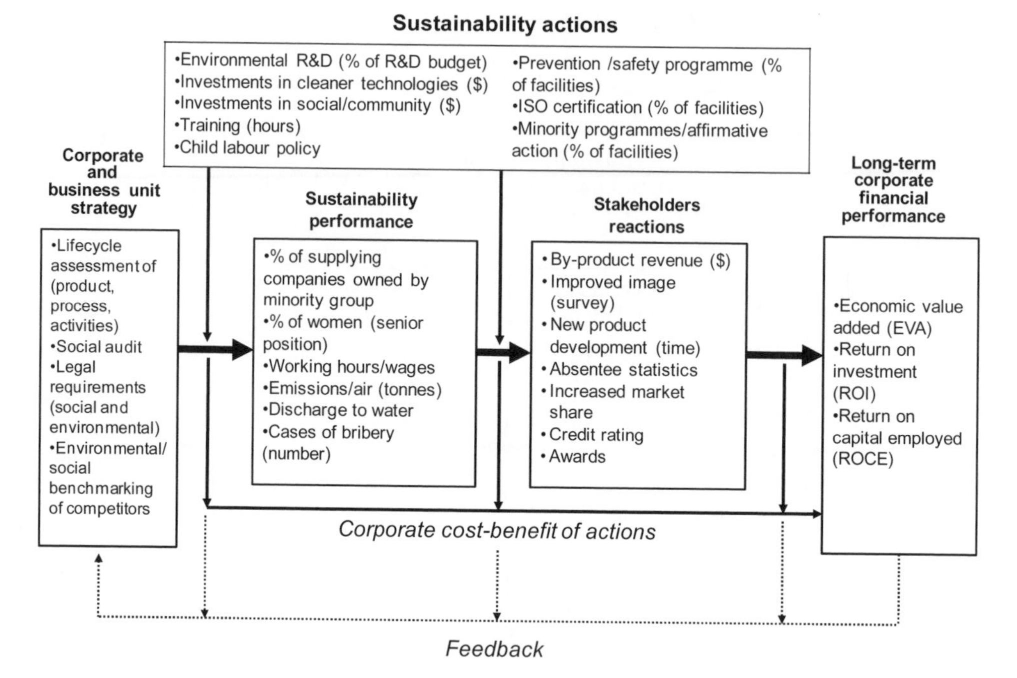

Source: Epstein and Roy (2001), "Sustainability in Action: Identifying and Measuring the Key Performance Drivers", *Long Range Planning*, Vol. 34.

In terms of comparability for external benchmarking, KPIs are not suitable because they are principally customised for each company based on system analysis in terms of mission, stakeholder expectations and goals. They would only be suitable for external benchmarking if a group of companies or an industrial sector with similar organisational structures, missions, stakeholders and strategies were to agree upon the characteristics of the KPIs to be used.

For applicability to SMEs, the preparation for organisational analysis might be an obstacle. In practice, the management of companies considering adoption of KPIs as a business tool sometimes find KPIs too expensive and the exact measurement of the performance required for a particular business or process objective too difficult. Since KPIs usually reflect long-term considerations, they may not suitable for SMEs, as they may need to modify structures, business models and target customers as well as strategies frequently. However, there is scope for developing simplified KPIs for SMEs to facilitate their understanding of the overall performance.

Box 3.2. Ford of Europe's Product Sustainability Index

Ford Motor Company Europe introduced new product design management as one way to tackle sustainability challenges such as climate change, oil dependency and air quality. This resulted in establishing the Ford of Europe's Product Sustainability Index (PSI), by which various dimensions of sustainability are combined into a comprehensive set of metrics for steering vehicle development. Since automotive product development needs very long lead times and changes take several years to trickle through to buy-in, cycle planning, kick-off, development and launch, product development in automotive industries has greater importance in management decision making than in other industries. Thus, the PSI has been carefully formulated to reflect the overall impact of the different vehicle attributes and makes the trade-offs visible (*e.g.* between life cycle global warming potential and the life cycle cost of ownership).

The PSI indicators are:

- life cycle global warming potential (greenhouse gas emissions along the life cycle);
- life cycle air quality potential [summer smog creation potential along the life cycle (volatile organic compounds, nitrogen oxides)];
- sustainable materials (use of recycled and natural materials);
- restricted substances;
- drive-by-exterior noise;
- safety (pedestrian and occupant);
- mobility capability (luggage compartment volume plus weighted number of seats related to vehicle size);
- life cycle ownership costs (vehicle price plus three-year fuel costs, maintenance costs, taxation and insurance minus residual value).

The PSI has been implemented with a process-driven approach. Clarification of the organisational context is of utmost importance in large and complex corporations in order to make individual departments directly responsible for the specific aspects of sustainability that can be affected by their area of responsibility.

Source: Schmidt (2008), "Developing a Product Sustainability Index", in *Measuring Sustainable Production*, OECD Sustainable Development Studies, OECD, Paris.

For management decision making, KPIs provide quantifiable milestones that reflect progress towards the organisation's goals, missions and stakeholders and information on its organisational structure and mechanisms. KPIs thus provide management with adequate information for their decision making in a long-term perspective.

In terms of improvement at the operational level, KPIs may not be effective as they use a restricted number of indicators selected to reflect key organisational challenges and lack operational level information. When operation-level indicators are part of KPIs, KPIs may apply to operational improvement, as illustrated by Ford of Europe (Box 3.2).

KPIs are not suitable for data aggregation and standardisation because they are customised for each company. Nor are they necessarily suitable for finding innovative products or solutions since their primary aim is strategic evaluation. However, if they include management indicators that set targets for innovative products/solutions (*e.g.* number of eco-labelled products), they could motivate employees to develop innovative ideas and put them on the market.

Composite indices – synthesising indicators to present a single message

Composite indices synthesise groups of quantitative and qualitative individual indicators to express a complex phenomenon through a limited number of indices. They are effective especially for presenting a large amount of information in an easily understandable format for management or external clients. They limit the number of statistics and serve as summary indices, and thus allow for ready interpretation and comparisons of relative performance.[4] The steps generally taken when structuring composite indices (OECD, 2003) are:

- develop a theoretical framework for the composite;
- identify and develop relevant variables;
- standardise variables to allow comparisons;
- weight variables and groups of variables;
- conduct sensitivity tests on the robustness of aggregated variables.

Figure 3.3 presents a model for composite indices for sustainability performance (Krajnc and Glavič, 2005a; 2005b). The calculation of the indices is a step-by-step procedure of grouping various basic indicators into sub-indices for each group of sustainability indicators. Sub-indices are combined into composite indices.

Figure 3.3. Structure of developing a composite index

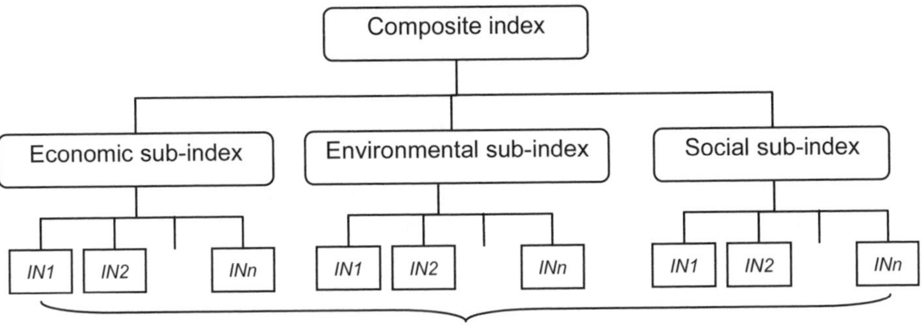

Source: Krajnc and Glavič (2005), "A Model for Integrated Assessment of Sustainable Development", *Resources, Conservation and Recycling*, Vol. 43.

The main issues in aggregating indicators are normalisation and weighting. Normalisation of each indicator is indispensable because indicators may be expressed in different units. Z-score, the most common method of normalisation, converts indicators to a common scale with a mean of zero and standard deviation of one. Appropriate weighting of indicators is also essential because it balances the significance of different sustainability attributes, taking into account a diversity of strategic emphases according to company and sector.

Composite indices can be suitable for external benchmarking as they give both simplified and quantified expressions of a more complex body of several indicators. They can be used to compare and rank companies within a specific sector. However, establishing composite indices usually requires careful consultation and negotiation among companies on the selection and weighting of objective indicators.

In terms of applicability to SMEs, the steps necessary for aggregating indicators could be an obstacle. Composite indices are more suitable for sectors that are able to convince their supply chain companies to adopt the same indicator sets. If appropriate software to facilitate the collection and processing of data were provided, this might encourage SMEs to adopt this approach.

For management decision making, composite indices can be useful because they simplify the information from a complex indicator set covering various aspects of corporate activity. Decision makers easily interpret composite indices and their sub-indices, if they do not have to identify a trend by studying many individual indicators. However, reducing the number of indicators by condensing information carries the risk of misinterpretation since users are not always aware of the scope and limitations of the indexing methodology and the message may be distorted by gaps in the data and by the way indicators are selected and weighed.

Regarding improvement at the operational level, composite indices can be applied effectively if they combine sufficient information on operations. For example, the sector-specific composite index presented in Box 3.3 was developed specifically to demonstrate the contribution of the steel industry to sustainable development in terms both of management decision-making and operational performance (Krajnc and Glavič, 2005b).

Box 3.3. The steel industry's Composite Sustainability Performance Index

Singh *et al.* (2007) evaluate the effectiveness of a composite index for the steel industry by using a case study from the Bhilai Steel Plant of Steel Authority of India Limited (SAIL). Apart from the three pillars of performance, organisational governance and technical aspects were considered as pillars for evaluating the sustainable performance of steel plants. A survey conducted by experts from different functional areas of the steel company identified a framework of Composite Sustainability Performance Index which combines 60 indicators from five categories. Its aim is to formulate a uniform methodology for assessing steel companies through comparison and thus effective decision making.

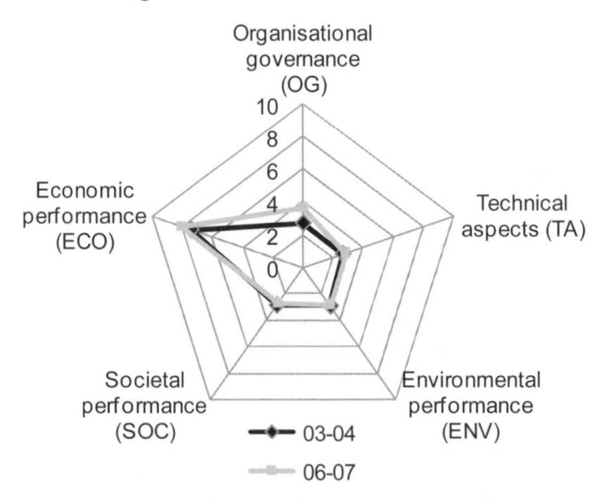

The overall score and sub-indices of various aspects of sustainability are evaluated by multiplying the global weights and adding the values of the respective aspects. These scores are normalised to 10 points based on the data collected for the company; the mean value of data is evaluated for each indicator. The actual values of different sub-indices for the evaluation year are plotted on the corresponding axes and the joining of points forms a new five-sided polygon.

Source: Singh (2008), "Developing a Composite Sustainability Index", in *Measuring Sustainable Production,* OECD, Paris.

Composite indices can facilitate data aggregation and standardisation within a sector once consensus has been reached. Ideally, they could show sector-level performance together with objective benchmarking of each company. However, use of the composite indices for data aggregation and standardisation beyond the sector is unlikely.

If appropriate indicators are selected, composite indices can be used to encourage the identification of innovative products or solutions. However, they cannot help develop individual products/solutions unless users return to the original indicators and sub-indices before aggregation. However, they can highlight opportunities for improvement and respond to emerging issues and pressures (Krajnc and Glavič, 2005b).

Material flow analysis – accounting for resource inputs and outputs

Worldwide, the use of virtually every significant material has been rising for many years, causing recurrent concerns over shortages in the stocks of natural resources, energy security and environmental impact (OECD, 2008a). Material flow analysis (MFA), a form of material balance analysis, aims to track the movement of materials from extraction to manufacturing, use in a product, reuse, recycling and eventual disposal, and to show effects on the environment at each step. MFA studies can focus on the whole economy, sectors, companies, or individual materials, products or substances.

MFA recognises that material throughput is required for all economic activities and asks whether the flow of materials is sustainable in terms of the environmental burden it creates. It accounts for all materials and energy used in production and consumption, including the hidden flows of materials that are extracted in the production cycle and do not enter the final product. The size of these hidden flows is often larger than the flows in the resulting products.

In essence, MFA has two main elements. First, material flow accounting, an accounting system for materials expressed quantitatively in physical units (tonnes, kilograms, etc.), describes the material flow as extraction, production, transformation, consumption and recycling, as well as disposal as waste or emissions to air or water (Peele, 2005). Material flow accounting includes inputs, outputs, and accumulations in material stocks. Second, material flow indicators derived from these accounts – such as direct material input, total material requirement and total material consumption – convey policy-relevant messages to a non-expert audience about the significance of material flows with respect to economic and environmental issues.

Within companies, the physical balance of inputs and outputs is increasingly used as part of environmental performance reports and provides substantial information for environmental management. MFA is useful for monitoring developments in resource productivity and environmental performance at the company or plant level. It also helps to set corporate strategies on investments and emissions and to monitor the availability of critical resources and the vulnerability of a company or a plant to disruptions in the supply chain. MFA of particular industrial materials, such as metals, can shed further light on concepts such as resource productivity and their relation to labour productivity, raw material prices and competitiveness (Box 3.4) (Bringezu, 2003).

Box 3.4. Material input per service unit

Material input per service unit (MIPS), which was originally developed by Germany's Wuppertal Institute in the 1990s, measures the total mass of material inputs to create a unit of service output. MIPS can be applied to whole economies, individual sectors of an economy or companies, as well as to single products and services or types of material or material groups by taking either a problem- or system-oriented approach. The MIPS methodology was designed to provide "a simple indicator of the material intensity of a product or service".

MIPS covers all material inputs at all phases of the life cycle of the product or service under investigation, including extraction of materials, manufacturing, transport, use, maintenance and end-of-life. The total mass of material inputs across the life cycle is aggregated to produce a single score for a particular product, and the score is represented per unit of service the product delivers. The results of a MIPS study can be used as a single indicator that represents material intensity across five categories: abiotic raw materials, biotic raw materials, soil movements in agriculture and forestry, water and air.

The strengths of the MIPS methodology include the comprehensive scope of material inputs across the product life cycle and the fact that it produces an easy-to-understand indicator. A shortcoming is that MIPS treats all materials equally and hence does not account for the qualities of material flows or environmental impact of different types of materials, their toxicity, transport or exposure pathways. It also does not consider the relative scarcity or abundance of materials.

Source: OECD (2007), "A Study on Methodologies Relevant to the OECD Approach on Sustainable Materials Management", OECD Environment Directorate.

The identification of waste is a major issue in MFA and allows for monitoring the waste typically unaccounted for in traditional economic analyses. It is thus a method for evaluating the efficiency with which material resources are used. Tracking the value of materials and their flow rates can show where value as well as material is lost (Box 3.5). MFA achieves this by using available production, consumption and trade data in combination with environment statistics, although it may not necessarily provide company-specific analysis.

Box 3.5. Material flow cost accounting

Material flow cost accounting (MFCA)[1] is a management tool for reducing the relative consumption of resources and material costs and can be applied in service industries as well as manufacturing industries. MFCA is a major tool of environmental management accountingand is oriented to internal use within an organisation.

MFCA enables the calculation and management of quantity and cost data for losses incurred in the manufacturing process. It views the final shipped products of the manufacturing process as "positive products", and emissions and waste along the way as "negative products". The material costs associated with negative products, processing and waste treatment costs are "negative product costs". Analysing the quantity of negative products and reducing the number of negative products makes it possible to reduce environmental burden and costs.

Canon, a Japanese camera and optical apparatus manufacturer, started using MFCA at a manufacturing line in a main factory for lenses. From the standpoint of MFCA, lens polishing sludge constituted a material loss. A joint MFCA project between Canon and its raw material suppliers was initiated in 2004, with both sides working together to reduce environmental burden and costs. As a result, a new thinner glass material was developed which reduces polishing sludge. Based on this success, Canon now deploys MFCA in the whole company. In 2006, Canon's environmental accounts show investment of JPY 19.1 billion in environmental protection, including JPY 5.8 billion for improvements designed to obtain economic benefits from environmental protection. This investment generated benefits of JPY 6.2 billion (Canon, 2007).

Japan's Ministry of Economy, Trade and Industry (METI) submitted the MFCA methodology to the International Organization for Standardization's technical committee on environmental management (ISO/TC 207) as a New Work Item Proposal (NWIP). In March 2008, the proposal was approved and ISO/TC207 Working Group 8 was set up to establish an ISO standard in three years' time.

1. MFCA can be considered as a hybrid of material flow analysis and environmental accounting.

Source: METI (2008), METI response to the OECD questionnaire on tools for sustainable manufacturing.

"Ecological footprint" is another variation of popular resource management tools. It uses input-output analysis to measure how much land and water a human population requires to produce the resources it consumes and to absorb its waste under prevailing technologies. At the company level, for example, SITA, a French waste management company, has created a tool for calculating the ecological footprint of the waste collection portion of their operations, and uses this to determine how to lower their ecological impact and increase the efficiency of their operations, as well as for communication with customers (Wackernagel, 2008).

In terms of comparability for external benchmarking, MFA is suitable as it is principally designed to provide aggregate background information on the composition of and changes in the physical structure of systems. Material-flow-based indicators can be aggregated from the micro level. One would need to set objectives for comparison and adjust organisational sizes and boundaries among companies or sectors.

MFA can be applied by SMEs, especially when the material balance of manufacturing procedures is analysed through basic metrics such as material input and waste generation. However, expert support may be needed to identify hidden flows of materials apart from tangible material flows through within a company.

MFA can be effective for management decision making as issues relating to materials have been increasing in significance for management owing to the rapid increases in the price of oil and raw materials over the last few years. MFA would be more useful for management decision making if it were combined with calculation of the cost of materials as this helps to identify where the company can cut costs (Box 3.5).

For improvement at the operational level, MFA can be very useful for identifying ways of minimising material inputs and outputs and thus make production processes most efficient.

MFA can be used effectively for data aggregation and standardisation, as it is mainly designed to provide aggregate information on the composition of and changes in physical structure. Company- or facility-level material flow accounting can be relatively easily compiled depending on the purpose for which the information is used. The basic data may be readily available from internal business sources. The major challenge is to ensure a minimum coherence with meso- and macro-level material flow accounting (OECD, 2008b).

MFA can be extensively used to find innovative products/solutions, as it helps to identify ways to minimise material inputs and outputs for making products/services. If appropriate benchmarks are available, MFA can help highlight opportunities for improvement and respond to emerging issues and pressures.

Environmental accounting – evaluating the profitability of environmental investment

Environmental accounting is based on a common financial accounting system. It is a systematic way to measure important environmental factors (Jónsdóttir *et al.*, 2005). At its simplest, environmental accounting makes environment-related costs more transparent in corporate accounting systems and reports. It is also a tool for evaluating the (economic and physical) effect of the cost (investment and expense) required or invested for environmental protection. Environmental accounting can be applied to the management of companies to link environmental issues with financial cost accounting and to evaluate the potential for "win-win" environmental protection and financial profitability. It is also applicable to accounting at the local and national levels.

The concept of environmental accounting was introduced around 1990 as a proactive approach to sustainable development. Its popularity has rapidly increased among companies in recent years, as identification and greater awareness of environment-related costs provide an opportunity to find ways to reduce or avoid these costs and improve environmental performance (Palme and Tillman, 2008).

It is important for management to uncover and recognise environmental costs associated with production. However, it may not always be clear whether a cost is "environmental" or not. The following are clearly environmental costs: costs incurred to comply with environmental regulations, costs of environmental remediation and pollution control equipment, and non-compliance penalties. Some costs fall into a gray zone or may be classified as partly environmental. For example, the costs of production equipment may be considered environmental if this equipment is considered part of a clean technology. The development of environmentally sound products/services might be also considered as part of environmental costs. Some companies even include the costs of environmental education, campaigns, donations and voluntary activities. It may also be difficult to distinguish environmental costs from health and safety costs or from risk management costs. Some governments provide national guidelines for corporate environmental accounting that help standardise what can be counted as environmental costs (*e.g.* MoE, 2005).

However, whether or not a cost is environmental may not be very important unless it is used when comparing one company to another, since the primary goal of environmental accounting is to ensure that relevant costs receive appropriate attention within a company. To handle costs in the gray zone, some firms use the following approaches (EPA, 1995):

- allowing a cost item to be treated as environmental for one purpose but not for another;
- treating part of the cost of an item or activity as environmental;
- treating costs as environmental for accounting purposes when a firm decides that more than 50% of the cost is considered "environmental" in nature.

Difficulties are greater when companies would like to estimate the economic benefits they can achieve from environmental investments. However, many benefits may be realised in the medium to long term, whereas available approaches tend to capture tangible short-term gains. Costs can be extended to include indirect ones borne by external parties such as consumers, communities and biodiversity by applying methodologies such as cost-benefit analysis and total cost assessment.

Environmental accounting can be used for external benchmarking if serious attention is given to comparability in composing the environmental accounting data. The guidelines developed by some governments may help companies provide consistent data by indicating what can be included as environmental costs and benefits.

SMEs can use environmental accounting as it is based on the existing framework of financial accounting usually adopted by SMEs. However, the initial cost of environmental accounting is relatively high and external help would be needed. Proactive entrepreneurs could take advantage of environmental accounting to significantly reduce environmental costs.

For management decision making, environmental accounting can be useful because it focuses on calculation of costs and gives results in simple monetary terms. Environmental accounting can provide management with useful data that take the environment into consideration and encourage continuous increases in environmental efforts. It can also be applied for decision making about investment in new process technologies and redesign of products/services.

At the operational level, environmental accounting can be effective because it focuses on environmental costs to be reduced or eliminated in operations, housekeeping and improvement of processes/products. By employing environmental accounting at one of its sites, the company can also obtain information to facilitate effective and efficient environmental activities aimed at resolving local environmental issues.

For data aggregation and standardisation, environmental accounting looks promising because it is based on financial accounting systems. In terms of international standardisation, environmental accounting at national level has been formalised into the System of Integrated Environmental and Economic Accounting (UN *et al.*, 2003), and some guidelines for environmental management accounting have been proposed (IFAC, 2005; UNDSD, 2001). However, the connection between national-level and corporate-level accounting systems is still weak.

Environmental accounting can be used to identify innovative products or solutions since it enables management to make pragmatic decisions on investment in innovative processes/products as the results are monitored in monetary terms. If appropriate benchmarks are available, it can work as an effective compass for eco-innovation path-finding.

Eco-efficiency indicators – identifying improvements in relation to economic value

Eco-efficiency indicators are quantitative indicators that specify the relation between economic value created and environmental impacts caused by the same institutional or geographical unit. They focus on the interplay between economic and environmental aspects and are in principle two-dimensional. They may be applied to a specific economic activity such as a production process, to a set of activities such as a product system, to a firm, to a sector, to a region or country, or to the global economy.

Use of the term eco-efficiency has been promoted through the activities of the World Business Council on Sustainable Development (WBCSD) since the early 1990s (Schmidheiny, 1992). The WBCSD defines eco-efficiency as "a management philosophy that encourages business to search for environmental improvements which yield parallel economic benefits" (WBCSD, 2000, p. 8).

Since eco-efficiency can be viewed from numerous perspectives and used on different levels, no single standard methodology for indicator systems has yet been developed. Two basic methodologies are used in eco-efficiency analysis: value-based eco-efficiency accounting and cost-based eco-efficiency accounting.

In value-based eco-efficiency accounting, the relation between economic value and environmental impacts is often summarised in an algebraic ratio, which either measures economic value created per unit of environmental impacts ("environmental productivity") or accounts for environmental impacts per unit of economic value ("environmental intensity"). The ratio of environmental productivity is the inverse of that of environmental intensity.

Cost-based methodology processes data in a similar way. Environmental improvement per unit of cost can be called "environmental cost-effectiveness". The inverse of this ratio conveys similar information and is called "environmental improvement cost" (*e.g.* marginal cost of emission reduction) (Huppes, 2007).

The concept of eco-efficiency is beginning to be applied in the daily operations of companies. Various manufacturing companies have developed in-house metrics for eco-efficiency (Figge and Hahn, 2004). These metrics allow managers to recognise at an early stage and systematically detect economic and environmental opportunities and risks in existing and future business activities. The concept of "Factor" is a practical application of eco-efficiency for environmental improvement of products/services and has been widely applied by Japanese electronics companies (see Box 3.6).[5] The formula of Factor is principally expressed as follows:

$$(Factor) = \frac{(Eco\text{-}efficiency\ of\ a\ product\ to\ be\ assessed)}{(Eco\text{-}efficiency\ of\ the\ benchmark\ product)}$$

Box 3.6. Application of an eco-efficiency indicator system at Panasonic

Panasonic, a Japanese electronics manufacturer, has been applying the concept of "Factor X" as an eco-efficiency indicator system "to quantify the way in which the product value can be increased while reducing the impact on the environment". By comparing the eco-efficiency of both new and old models of a product, the level of improvement is expressed in the number of times greater the eco-efficiency of the new model is than that of the old model.

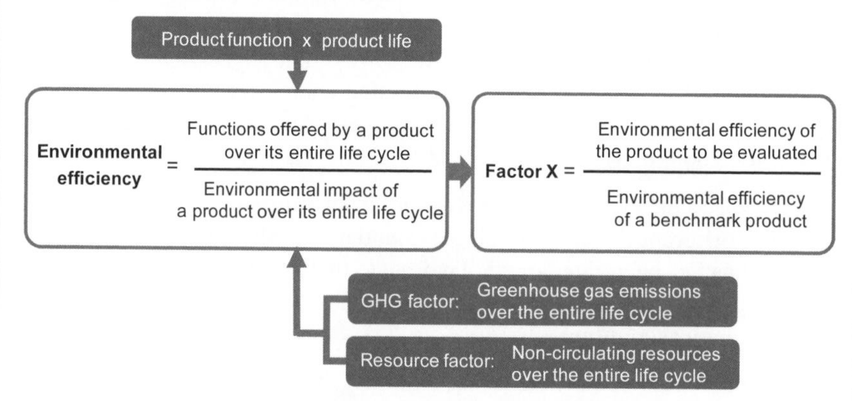

Panasonic applies Factor X to two major environmental aspects – greenhouse gas (GHG) emission reduction and efficient resource use. The GHG factor and the resource factor are defined as in the chart above. Factor X is expressed using simple mathematical values to indicate the level of improvement in these eco-efficiency criteria and utilises these values in subsequent evaluations or as numerical targets in product development.

In this way, Panasonic evaluates whether or not adequate efforts are being made to minimise environmental pollution risks throughout the production and distribution system.

Source: Panasonic website, *http://panasonic.net/eco/products/factor_x*.

Eco-efficiency indicators can be put into operation in a number of ways. Because of the expansion of eco-efficiency as a conceptual and operational framework, functional comparability of eco-efficiency in terms of company performance is not yet possible. Companies continue to develop eco-efficiency analysis in-house or publish eco-efficiency indicators on a voluntary basis as part of sustainability reporting. If the methodology and the industry baseline were unified, eco-efficiency indicators could become a very powerful tool to encourage sound competition in providing more efficient products/ services and processes through external benchmarking.

Technically, eco-efficiency indicators can be implemented by SMEs as a unit of accounting and reporting. Nevertheless, in practice, SMEs lack the necessary managerial and financial resources and awareness, or they have no incentive to adopt such advanced indicator systems.

In principle, eco-efficiency indicators can be applied to products/services and production processes, as well as to overall corporate performance. However, most applications are found at the level of operations.

Eco-efficiency indicators collect information that can be used in daily company operations and decisions. Some companies that have developed in-house metrics have started to integrate eco-efficiency estimates into their operational management. This shows that eco-efficiency indicators can be used to achieve incremental cost gains and deliver more long-run economic value. They may provide operational managers with the possibility to detect systematically and recognise at an early stage economic and environmental opportunities and risks in existing and future business activities.

Eco-efficiency indicators can be aggregated and standardised in a number of ways depending on their conceptual, institutional and operational context. However, most applications are found at the level of products/services and production processes.

Eco-efficiency indicators can generally support incremental innovation in products and processes and could potentially facilitate more radical innovation when being applied at the company level.

Life cycle assessment – embracing cradle-to-grave management

Life cycle assessment (LCA) is defined as a study of "the environmental aspects and potential impacts of a product or process or service throughout its life, from raw material acquisition through production, use and disposal" (ISO, 1997). The term refers to the evaluation of the entire life cycle of a product, "from the cradle to the grave", *i.e.* from the extraction of basic resources, through production and transport, to use and disposal of the product itself. The LCA methodology can address both quantitative and qualitative aspects of a single product, a material or a group of materials, as well as services from the life cycle perspective.

LCA is often used to compare products with equivalent functions or to determine “hot spots” during the life cycle which are critical to the overall environmental impact. For a specific product, one only sees a small part of the total material flows mobilised in the course of its production. The “hidden” flows, such as fossil fuels used in manufacturing and transport, should be considered part of the product’s total impact on the environment. LCA can help companies identify important aspects of the production process from the sustainability perspective.

An internationally standardised method of LCA was developed as the ISO 14040 series. These standards advise companies to carry out LCA in four distinct phases: *i)* defining goal and scope; *ii)* making a life cycle inventory (ISO 14041); *iii)* conducing life cycle impact assessment (ISO 14042); and *iv)*interpreting the assessment results (ISO 14043).

LCA also provides a wide range of environmental tools that incorporate life cycle thinking. It allows for an analysis of problems related to a particular product/service, for comparing improvement variants of a given product/service, for designing new products/services, and for choosing among several comparable products/services. Eco-design is one approach to assessing the environmental aspects of a product/service and is often based on LCA. Eco-design aims to ensure that new products/services are designed to cause minimal environmental damage over their life cycle.

LCA can also help individual and institutional consumers to make purchasing decisions. Eco-labels have been widely applied to products as a way to communicate their life cycle environmental impact, as calculated by LCA, to consumers and to make it easier for them to choose more environmentally sound products/services.

The “carbon footprint” is a recent use of LCA which aims to make production more sustainable (Box 3.7). It may be defined as a measure of the total amount of carbon dioxide (CO_2) emissions directly or indirectly caused by the activity or accumulated over the life of the product.

LCA results are comparable in principle as the methodology led to the international standards of ISO 14040-44 enables the comparison of environmental impacts over the life cycle of material use and associated emissions and energy requirements. The results of LCA can be presented in common comparable units. In practice, however, the fact that users of LCA data tend to make different assumptions and set system boundaries to fit their individual needs has made it difficult to compare similar products produced by different companies.

Box 3.7. Carbon footprint

"Carbon footprint" has been applied as an eco-label to a range of products to indicate the CO_2 emissions generated throughout the product's life cycle. A number of approaches have been proposed to provide estimates, ranging from basic online calculators to sophisticated LCA or input-output-based methods and tools. The concept not only enables companies to demonstrate their efforts to reduce CO_2 emissions, it but also improves consumer awareness of the issue.

The British Standards Institution (BSI) is currently leading the development of a Publicly Available Specification (PAS) to measure the embodied GHG emissions from goods and services across their life cycle. The method was developed by the Carbon Trust, an independent company set up by the UK government. The PAS 2050 was launched in October 2008.

Japan's Ministry of Economy, Trade and Industry (METI) has convened a study group to research possible programmes and methodology for carbon footprint. Environmental managers from over ten companies, including manufacturers, are participating. The METI intends to establish guidelines for calculating and displaying carbon footprints and is proposing the standardisation of carbon footprint to the ISO.

Source: Carbon Trust website *www.carbontrust.co.uk/carbon/briefing/pre-measurement.htm*; and METI (2008a), METI response to the OECD questionnaire on tools for sustainable manufacturing.

LCA can be used by SMEs because various software tools for application are available. These tools can also help them provide the database for life cycle inventory, the most difficult obstacle to LCA use. However, the use of LCA has generally been considered to be too resource-intensive for SMEs.

Relatively simple expressions of LCA results also serve to inform management about indirect environmental effects of companies' operations beyond their organisational boundaries and hence encourage more systemic thinking. However, LCA is only applicable at the level of products/services and not the entire company.

LCA is also capable to improve operations as a comparison of LCA results makes it easier to identify which parts of the production processes need to be improved. It is possible to identify the stages of the production process with the highest environmental impact and thus improve them.

For data aggregation and standardisation, LCA data can be aggregated and standardised if their results are presented in common comparable units such as kg CO_2-equivalent. Once consensus on system boundaries and expression of results is reached among all concerned parties, LCA can be used for data aggregation and standardisation on a product/service basis.

LCA has strong potential for identifying innovative products/solutions within the cradle-to-grave scope. Because LCA provides information about impacts of a product over its life cycle, companies can evaluate new processes/products in a holistic manner. LCA may be used for evaluating the feasibility of potential products in terms of environmental impact by testing prototypes or through simulation.

While LCA is thus a strong tool to provide life cycle thinking, it has faced a number of challenges, including difficulties in setting consistent system boundaries for fair comparisons, data reliability and quality of life cycle inventory, and consistent weighting of the data in impact assessment. These challenges need to be tackled in order to help consumers to make good decisions on choosing environmentally friendly products/services and eventually to enable companies to benchmark their sustainable production initiatives against those of others based on LCA. The challenges are greater for sectors with complex supply chains (Hauschild *et al.*, 2005). Further standardisation of the LCA methodology is essential to enable meaningful evaluation and comparison.

Sustainability reporting indicators – informing stakeholders about activities and progress

Sustainability reporting indicators are a set of indicators which organisations can use to disclose information about the performance of the economic, environmental and social aspects of their activities and processes. It can be applied to a variety of institutional or geographical units at various levels, but has been mostly used at the facility, company and sectoral levels.

Early models of sustainable reporting indicators can be found in the environmental reporting initiatives of chemical companies which suffered from serious image problems in the late 1980s.[6] Today, companies can use them to identify and manage non-financial and intangible risks and opportunities connected to their operations through measurement and data collection. An increasing number of governmental departments and local authorities also publish sustainability reports.

Governments in Denmark, the Netherlands and Portugal have made sustainable reporting mandatory for public agencies and private companies. There are even efforts to mainstream sustainability reporting by requesting non-financial disclosure as part of mandatory annual financial accounts, as in France's new economic regulations. Australia, Austria and Japan are among those taking a voluntary approach by providing guidelines that standardise sustainable reporting indicators. However, the Global Reporting Initiative's (GRI) Sustainability Reporting Guidelines are rapidly becoming the internationally accepted voluntary framework for sustainability reporting used by companies around the world (Box 3.8). These sustainability reporting indicators combine quantitative and qualitative information. They are often categorised in terms of the three pillars of sustainable development. Most guidelines also ask for information on the organisation's mission, governance and management system relating to sustainability. While the perspective on sustainability is multi-dimensional, all indicators are independent.

Box 3.8. The Global Reporting Initiative

The Global Reporting Initiative (GRI) was established in 1997 by the Boston-based Coalition for Environmentally Responsible Economies (CERES), with the vision that "reporting on economic, environmental, social performance by all organisations becomes as routine and comparable as financial reporting". It soon became a multi-stakeholder international organisation with support from the United Nations Environment Programme (UNEP).

The work of the GRI is targeted at companies and other organisations of all sectors and sizes interested in reporting sustainability aspects of their activities. In addition to its Sustainability Reporting Guidelines, which are generally applicable to all businesses, supplements provide guidance for particular sectors. The GRI Guidelines are the result of a continuous process of consultation with its stakeholders such as organisations applying the guidelines and other experts.

The guidelines have three parts. The first part contains the principles of sustainability reporting with regard to the content and scope of the report. The reporting organisation is expected to develop its sustainability reports based on certain principles including relevance, completeness, comparability, accuracy and transparency. The second part provides a list of relevant indicators on the economic, environmental and social performance of the company or organisation. The third part contains advice on more general questions such as how to use the guidelines and how to ensure the credibility of a report. The guidelines list 13 indicators for economic performance including economic value generated and spending on locally based suppliers, 35 indicators covering the environmental performance of the organisation in terms of water, energy, biodiversity and other important environmental media, and 49 social indicators cover statements about management practice and child labour as well as corruption and community involvement.

Over 1 000 organisations from 54 countries issued their sustainability reports based on the GRI Guidelines over 2008.

Source: GRI website *www.globalreporting.org*.

Although sustainable reporting indicators were primarily developed for external disclosure, sustainability reporting is a way for companies to start collecting environmental and social data and monitor progress in order to improve their sustainability performance at the site and company levels. While sustainability reporting is still practised mainly by relatively large companies, the GRI issues a handbook to encourage SMEs to report their sustainability performance. It is also developing a series of sector-specific indicators to supplement the general set of indicators to respond to demands from industry, while providing technical protocols that aim to unify the measurement units and methodology as well as organisational boundaries.

The listing requirements for greater accountability and disclosure on corporate governance from stock markets and financial regulators are another good example of a context in which sustainability reporting is taking place. The business community also may affect the need and requirements for sustainability reporting through membership rules. A good example is the sustainability development framework for member companies of the International Council on Mining and Metals (ICMM). The UN Global Compact now requires signatories to report their sustainability performance annually using its ten universally accepted principles in the areas of human rights, labour, the environment and anti-corruption. Some sector associations such as chemicals, steel and aluminium provide their own reporting indicator sets and guidelines and compile the data reported from member companies (*e.g.* CEFIC, 2007; IISI, 2005; EAA, 2006).

Comparability is included in the reporting principles of the GRI Guidelines, and software providers offer standardised data processing options for sustainability reporting. However, comparability of data between reporting companies has not yet been achieved, partly because of the voluntary nature of sustainability reporting, the many qualitative indicators and the difficulty for setting consistent organisational boundaries.

Sustainability reporting frameworks offer ways to facilitate sustainability reporting by SMEs. Most guidelines are provided free of charge, and SMEs benefit from information services provided by non-profit platforms, public agencies or global initiatives. The GRI has a handbook for SMEs, but SMEs have undertaken relatively little sustainability reporting.

Sustainability reporting enables companies to present their overall vision and strategy for managing the challenges associated with economic, environmental and social performance. A quality report can show stakeholders and investors the measures the company is taking to reduce risks and seize opportunities. Thus, sustainability reporting can be an important tool for managing a company's decisions and operations in a more strategic and long-term perspective. The publication of sustainability reports can facilitate the integration of sustainability issues into mainstream management as strong commitment by management is indispensable for such disclosure.

Many sustainability reporting frameworks are sufficiently developed to cover the overall operational management of companies. Their timeliness, completeness and balance of information can enable companies to measure the sequence and timing of their activities. However, the presentation of individual indicators in sustainability reports does not necessarily help companies prioritise particular areas or consider alternatives in an integrated manner in order to improve their overall environmental performance. The GRI Guidelines recommend identifying "material" issues through stakeholder engagement rather than asking companies to report on all indicators in the guidelines.

In order to make data aggregation and standardisation possible, consistent organisational boundaries need to be set to avoid double counting. Although the GRI provides a boundary protocol, it does not set boundaries in as strict a manner as financial accounting. Sustainability reporting indicators also include qualitative data which are not suitable for aggregation.

Sustainability reporting itself does not help to find innovative products or solutions, as its aim is to provide information on corporate performance. However, as many reports also include information on products/services and production processes, they may indirectly help to improve production.

Socially responsible investment indices – benchmarking performance for financial markets

Socially responsible investment (SRI) refers to an investment strategy that seeks to maximise simultaneously financial return and social and environmental good. SRI indices are generic, generally composite indices which incorporate a number of quantitative and qualitative indicators. The approaches and methodologies reflect the criteria of investors in the growing SRI market in terms of economic, environmental or social sustainability. SRI indices aim to analyse and evaluate companies or industries for particular groups of financial investors, according to predefined criteria. Some leading banks also publish sustainability criteria which borrowers are required to meet for the financing of certain projects.

Thanks to the participation of institutional investors such as insurance companies, pension funds, and religious and other mission-driven associations, SRI has become a booming financial market in OECD economies. Assets in socially screened portfolios climbed to USD 2.71 trillion in 2007 in the United States, for a share of some 11% of professionally managed capital services (Social Investment Forum, 2008). The European SRI market grew to EUR 1.6 trillion in 2007 (Celent, 2007).

Box 3.9. Dow Jones Sustainability Indexes

The Dow Jones Sustainability Indexes (DJSI) have covered the performance of leading companies worldwide since its launch in 1999. Dow Jones Indexes, together with STOXX Ltd. and Sustainable Asset Management (SAM), provides benchmarks at different regional levels including global, European, North American, Asia Pacific and United States indices.

To construct a composite sustainability index, corporate sustainability criteria are initially identified by assessing economic, environmental and social driving forces and trends. Sustainability criteria can be either general or industry-specific. All criteria are based on widely accepted accounting, statistical and information standards and procedures. Weightings are attached accordingly.

To gather input, four major sources of information are used: company questionnaire, company documentation, media and stakeholders, and direct contact with companies. Finally, a company's total corporate sustainability score is calculated based on a predefined scoring and weighting structure.

Dow Jones Sustainability World Index's corporate sustainability assessment criteria and weightings

Dimension	Criteria	Weighting (%)
Economic	Corporate governance	6.0
	Risk and crisis management	6.0
	Codes of conduct/compliance/ corruption and bribery	5.5
	Industry-specific criteria	Depends on industry
Environment	Environmental performance (eco-efficiency)	7.0
	Environmental reporting*	3.0
	Industry-specific criteria	Depends on industry
Social	Human capital development	5.5
	Talent attraction and retention	5.5
	Labor practice indicators	5.0
	Corporate citizenship/philanthropy	3.5
	Social reporting*	3.0
	Industry-specific criteria	Depends on industry

* Criteria assessed based on publicly available information only.

Source: Dow Jones Indexes and Sustainable Asset Management (SAM) (2009), *Sustainability World Index Guide Book* (version 11.1), September, SAM, Zurich.

In earlier periods, SRI indices were simply based on negative selection criteria, *i.e.* investment was avoided in undesirable sectors such as tobacco, gambling, slavery and defence industry ("negative screening"). In recent years, a new approach looks for best practices among competitors to encourage companies to improve their performance through benchmarking ("positive screening"). Many investors now consider climate change as one of the most significant business and investment risks (Richardson, 2008).

Sustainability criteria have been arranged mainly by the leading financial index providers such as Dow Jones Indexes and FTSE and specialised rating agencies (Box 3.9). Financial firms and institutional investors either develop their own criteria or purchase rating information from these providers in order to make their decisions. As of October 2004, there were at least 12 "families" of market indices of sustainable companies, and over 35 individual indices in at least seven countries (Hamner, 2005).

SRI criteria are likely to have a strong influence on the sustainability aspects and practices companies need to focus on because they are regularly surveyed by rating agencies, and because the results of benchmarking are clearly comparable between competitors and directly influence investors' decisions. On the other hand, SMEs have not been a part of the growing SRI trend, since most investors focus on global and national companies. As some banks in OECD countries have introduced screening based on sustainability criteria for lending to SMEs, pressure from ethical investors may affect SMEs.

SRI indices also provide financial firms' evaluation of companies' strategies and management of sustainability opportunities, risks and costs. By integrating economic, environmental and social factors in their business strategies, companies can be motivated to focus on long-term shareholder and stakeholder value.

However, since SRI indices are set by external parties, they are not directly used by companies to improve their manufacturing processes and products/services at the operational level. Nor is it intended for data aggregation and standardisation. As each rating institution promotes its benchmarking criteria, establishing a unified approach is difficult. Also they cannot help companies identify innovative products or solutions unless the criteria include targets for innovative products/solutions (*e.g.* eco-labelled products).

Benchmarking indicator sets: key findings from the analysis

It is not easy to compare the various categories of indicator sets since their structure and scope of application differ. However, it is useful to try to establish, on the basis of the criteria used to analyse them, the context in which the indicator sets are most effective for advancing sustainable manufacturing. Table 3.3 summarises the results.

Table 3.3. Summary of the review of sustainable manufacturing indicator sets

Type of indicator sets \ Criteria	Comparability	Applicability for SMEs	Management decision making	Operational performance improvement	Data aggregation and standardisation	Finding innovative products or solution
Individual indicators	*	***	*	**	*	*
Key performance indicators	*	*	***	*	*	*
Composite indices	**		**	*	**	*
Material flow analysis	*	*	*	***	**	***
Environmental accounting	**	*	**	***	**	**
Eco-efficiency indicators	**	*	**	***	**	***
Life cycle assessment	**	*	*	***	**	***
Sustainability reporting indicators	*	**	**	**	*	*
Socially responsible investment indices	**		**			*

***: Strongly suitable for the purpose.
**: Suitable if certain conditions are met.
*: May be applicable but not necessarily suitable.

Note: The usefulness of each indicator set may also be subject to the competence and context of the applying organisation.

Comparability for external benchmarking

LCA has advantages compared with the other categories of indicator sets in terms of comparability because the methodology has been established as an international standard. However, they are mostly used for individual products/services. Also, application of the LCA methodology is not necessarily consistent so that comparisons between products/services of different companies are difficult. Even though eco-efficiency indicators lag in terms of comparability, standardisation to allow comparison may become possible. Composite indices appear suitable for external benchmarking as they rely on a limited number of figures, but companies or sectors still need to agree on the methodology. Environmental accounting also looks suitable, but further development of methodology and agreement on what counts as an environmental cost are needed. Although SRI indices were originally developed for external benchmarking, the number of companies involved so far is limited. Individual indicators can be used for benchmarking if companies agree on a core set of indicators for comparison.

Applicability to SMEs

Individual indicators are most commonly used by SMEs since they can be applied without special preparation. Although it can be resource-intensive, the use of LCA may be attractive because various support tools are available. SMEs might also use environmental accounting for sustainability assessment. The methodology is attractive because it might help reduce environmental costs while increasing economic benefits. Since sustainability reporting indicators provide a menu of well-designed indicators, they can be useful for SMEs willing to measure their performance for the first time. MFA and eco-efficiency indicators can technically be implemented by SMEs. KPIs and composite indices require preliminary procedures before they can be used.

Usefulness for management decision making

KPIs and composite indices are ideal for use by management. They are designed to assist decision making, although composite indices risk losing some useful detail. Environmental accounting can be useful because it is based on financial accounting. Economic valuation would encourage management to think about environmental investments not simply as costs but also as revenue-generating opportunities. SRI indices can provide management with good external benchmarks on their sustainability strategies and a better understanding of the opportunities, risks and costs of sustainability. LCA can help management identify hotspots for environmental efforts and encourage more systemic and value chain-based thinking beyond organisational boundaries. The significance of MFA and eco-efficiency indicators for management has been increasing owing to the high price of oil and other material inputs. Individual indicators can assist decision making if a restricted number of appropriate indicators with sufficiently relevant information are selected and presented as a scoreboard.

Effectiveness for improvement at the operational level

Environmental accounting is one of the most useful measures for reducing the costs of business operations while achieving the most benefit from environmental investments. LCA offers the best solution for reducing environmental impacts as a result of actual operational improvement throughout the value chain. MFA and eco-efficiency indicators can also be very useful for identifying ways to make resource use more efficient. The advantage of eco-efficiency indicators is that the same indicator sets can be used for both operational (product/services and processes) and management (corporate-level) improvements. Individual indicators and sustainability reporting indicators could be applied effectively if they contained enough information that is relevant to operations and if the indicators are compiled in a way that allow companies to understand how changes in products and processes affect individual indicators. The usefulness of composite indices and KPIs depend on whether operational-level indicators were set to be part of them and meaningfully linked with other managerial aspects.

Possibility for data aggregation and standardisation

MFA is principally designed to provide aggregate information. It would be good to standardise the methodology. The availability of company-specific data could be a problem. LCA datacan be used for aggregation since they can be presented in common comparable units, but require consensus among concerned parties regarding system boundaries and expression of results. Possibilities for further standardisation have been considered. Composite indices are also suitable owing to the relatively small number of indicators, but because they are company or sector-specific they are not useful for standardisation. Environmental accounting can be utilised for data aggregation if the definition of environmental costs and the methodology for identifying benefits are unified and standardised. Eco-efficiency indicators can also be standardised, but their effectiveness depends on the conceptual, institutional and operational context.

Effectiveness for finding innovative products/solutions

MFA, environmental accounting, eco-efficiency indicators and LCA can be useful for identifying innovative products or solutions. It is hard to judge which is best as they focus on different aspects of environmental and economic solutions. Composite indices and KPIs can be used for this purpose if appropriate indicators are selected as a benchmark for innovative products/solutions, but they can only target improvement in single environmental aspects or the whole range of products (*e.g.* number of eco-labelled products). Individual indicators could be helpful when links between indicators are established and observed. But generally there is risk of missing possible trade-offs between different environmental impacts by focusing on only a few independent aspects.

To sum up, the key findings from this benchmarking analysis are as follows:

- **Demand for information is increasing.** Manufacturing companies are operating under increasing pressure for better information on the sustainability of their products/services, activities and business strategies from government, investors and civil society. This pressure has been the main driving force behind models of sustainability measurement and management. There is growing acceptance among companies that sustainability measurement can lead to better informed strategies and more responsive customer service, in addition to better operational environmental performance.
- **Consistent measurement is a challenge.** The concept of sustainable development poses a significant challenge for measurement at the company level. The demand for information is varied, changes over time and originates from diverse sources such as company management, investors, communities and customers. It is critical for companies to choose the right methodology and the right elements to measure to advance sustainable manufacturing effectively. This is a challenge given the current proliferation of indicator sets. Conceptual approaches and operational frameworks used to implement sustainable manufacturing remain fragmented.
- **SMEs need to build capacity for measurement.** Increased competitive pressures due to technological shifts and globalisation are forcing companies to reconfigure their value chains. The production process is now diffused in a web of companies of different sizes in different locations. As the scope of sustainable manufacturing is also expanding from a single facility or company to "cradle to grave" or even "cradle to cradle", the engagement of supply chain and downstream companies, often SMEs, is becoming inevitable. However, most SMEs lack incentives to implement sustainability indicators and face structural bottlenecks and capacity gaps. It is strongly advisable for large companies and government to provide a range of supportive measures for increasing the use of indicators among SMEs.

- **There is no ideal indicator set.** Ideally, sustainability indicators should be able to serve two main purposes – management decision making and improvement in products/services and production processes. With the exception of eco-efficiency indicators, each of the nine categories of indicator sets is mainly designed either to help decision making by management or to facilitate improvements at the operational level. Each category has strengths and shortcomings and there appears to be no single ideal indicator set.

How are manufacturers applying indicators?

To acquire a better view of how indicators are actually used to advance sustainable production in manufacturing companies, the OECD conducted a questionnaire survey to companies. The survey was conducted between July and September 2008, and the questionnaire was sent to manufacturing companies through the Business and Industry Advisory Committee to the OECD (BIAC) and members of this project's Advisory Expert Group. The OECD received 40 responses mainly from the electronics, automotive and chemical sectors (Figure 3.4). In addition, a series of focus group meetings of corporate experts from the electronics and automotive/transport sectors were organised to obtain further direct input in September 2008 in Rochester, NY, and in November 2008 in Brussels. The following section presents the results obtained from these activities.

Current use of indicators

No single set of sustainability indicators is used by all companies, and their usefulness depends on the nature of products/services and manufacturing processes. Many companies are using more than one set of indicators at the same time, and they are often not comparable between different businesses and sectors.

The most widely applied indicator sets across industries appear to be those that are easy to use and adaptable to individual company situations and purposes. These include the compilation of individual environmental data (used by 88% of the survey respondents) and KPIs (used by 80%). These were also generally judged to be the most useful by the respondents because they are easily adapted to the objectives of each business. In the focus group meetings, this was reflected in a strong emphasis on the importance of keeping indicators simple and transparent, especially when used for external communication and benchmarking purposes. To this end, survey respondents also reported extensive use of reporting indicators, citing the GRI Guidelines. Overall, 73% of respondents participated in some form of corporate reporting scheme.

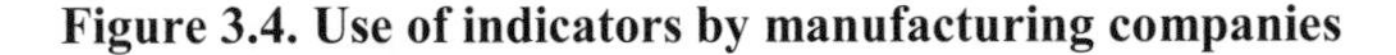

Figure 3.4. Use of indicators by manufacturing companies

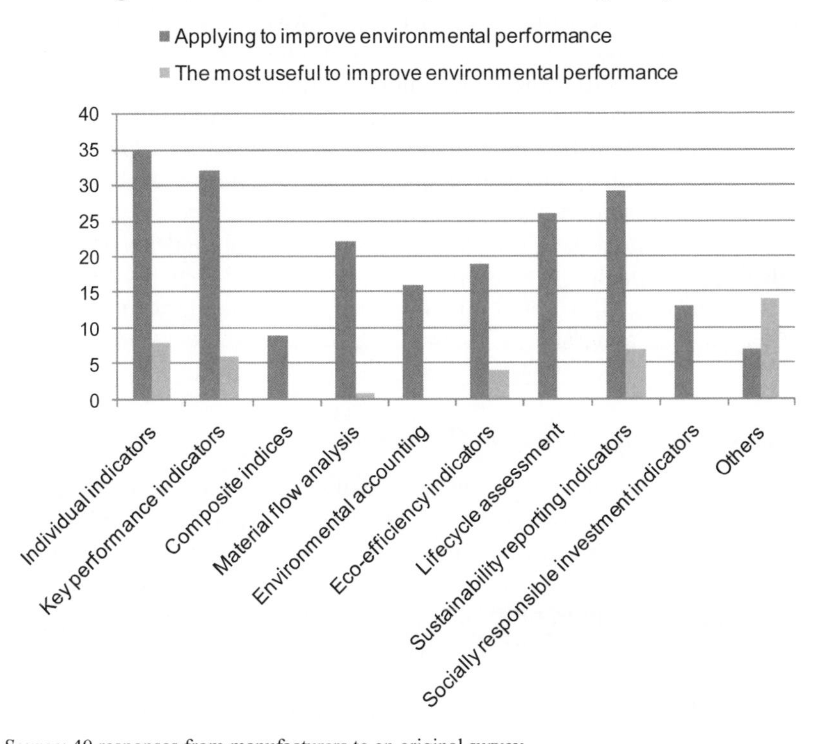

Source: 40 responses from manufacturers to an original survey.

For internal use to improve manufacturing processes and products/ services, however, indicators based on more complex methodologies such as MFA, eco-efficiency and LCA were considered very useful because they help to understand and manage the company's specific performance. A number of survey respondents reported the use of these indicators, predominantly LCA (used by 65%) and MFA (used by 55%, mainly from the chemicals industry). Those who found these useful generally viewed MFA and LCA as well-developed, internationally recognised methods which go beyond individual impact assessment and hence support a more far-reaching and systematic advance towards sustainable manufacturing. But survey respondents, as well as focus group participants, also pointed out that these advanced methods may not be easily applied in SMEs or by companies lacking experience in using indicators.

In the focus groups, many participants were concerned that the LCA methodology was too data-intensive. They also felt that users of LCA information might easily be confused owing to differences among companies in weighting, scoping and data sources. They indicated the need for simpler life cycle indicators as well as more transparency regarding the assumptions and data sources companies use for LCA. It was expressed that material flow information should become more important in order to consider resource efficiency from the viewpoint of increasing material scarcity and rising material costs. It was also suggested that the focus of environmental impact should be holistic rather than single aspects such as CO_2 emissions and energy use. Participants also emphasised that the further development of environmental valuation techniques, including environmental accounting, needs to be explored as a way to encourage more rational investments in sustainable manufacturing activities. It was proposed that resource-based indicators such as MFA could possibly integrate economic valuation based on potential costs and risks.

Barriers to the adoption of indicators

Survey respondents designated complexity as the main barrier to the use of sustainability indicators in their production, particularly by SMEs and companies with little experience in assessing their sustainability performance. Elaborating on these concerns, focus group participants drew attention to the lack of clarity on what to measure, how to measure, and how to compile or gain access to the necessary data while ensuring a certain level of data quality. The latter was of particular concern for large companies and those relying on a large number of subcontractors in their supply chain.

A related area of concern is the lack of comparability of products, processes, companies and sectors. Many companies often do not know where they stand or how far they have progressed as compared to competitors. Given these uncertainties, the costs of developing their own indicators may be seen as a considerable barrier to the application of sustainability indicators. Many companies also pointed out that businesses could not be expected to tackle such issues by themselves. Focus group participants, particularly from the electronics industry, mentioned that rapid technological changes could impede the application of sustainability indicators because there would not be enough time to develop and adopt relevant metrics.

The role of the government and the OECD

The results from both the survey and the industry focus groups demonstrate that there is little interest in the development of a new set of indicators and that governments and the OECD should instead look towards bringing clarity and consistency to existing indicators. The cross-sectoral consensus arising from the survey appears to be that existing indicator sets are either too complex and not comparable or basically sufficient to cover most business needs. The need for the harmonisation and simplification of indicators, and their promotion, was echoed in the industry focus group meetings as an area in which governments and the OECD could play a vital role. Both the survey respondents and the focus group participants also highlighted that such efforts should preferably be directed towards the mapping of existing indicators, the development of common terminology and standard measurement methodologies, and the provision of supportive tools.

Conclusions

This chapter reviews existing sets of indicators that assist industry and companies to track and benchmark different aspects of their performance in order to improve their production processes and products/services with a view to sustainable development. There is a multitude of such indicator sets, but they have not been comprehensively categorised and analysed. They are here classified into nine heuristic categories: *i)* individual indicators; *ii)* key performance indicators (KPIs); *iii)* composite indices; *iv)* material flow analysis (MFA); *v)* environmental accounting; *vi)* eco-efficiency indicators; *vii)* life cycle assessment (LCA); *viii)* sustainability reporting indicators; and *ix)* socially responsible investment (SRI) indices. The effectiveness of these sets of indicators is examined on the basis of predefined criteria.

As analysed above, and as indicated by both survey respondents and focus group participants, no single set of indicators in the nine categories covers everything manufacturing companies need to address to improve their production processes and products/services. A combination of indicator sets can instead help companies to obtain the most comprehensive and appropriate picture of economic and environmental impacts throughout their value chain and the life cycle of their products. The development of sustainable manufacturing indicators can be a continuous, evolutionary process of setting goals and performance measurement.

For example, it could be valuable to consider combining MFA, LCA and environmental accounting. MFA results alone can only show the physical figures of material flow through the economy (*e.g.* the entire company), but this could be complemented with LCA methodology to incorporate the product life cycle perspective. The use of environmental accounting would further strengthen the understanding of links between material use, financial implications and environmental impact. However, when used for management decision making and external communication, indicators need be simple and transparent.

Box 3.10. Development of an "environmental contribution indicator" in Japan

The potential contribution of information and communication technologies (ICTs) to tackling global environmental challenges has recently started to attract greater attention from industry and policy makers. In its latest report, the Climate Group, a UK-based non-profit organisation, estimated that the ICT sector currently contributes around 2% of annual global man-made CO_2 emissions, and the figure will almost double by 2020. But changes in the way people live and businesses operate through effective use of ICTs could reduce global CO_2 emissions by 15% during the same period. This opportunity for environmental contributions will be realised through smart ICT applications for building design and use, smart logistics, smart electricity grids and industrial monitor systems, as well as the replacement of physical products and services with their virtual equivalents, such as tele-working, video-conferencing and e-commerce (The Climate Group, 2008).

However, if measurement of environmental impacts focuses only on a single company or a single product, system-wide contributions may be missed. To balance out these negative and positive impacts, Japan's Green IT Initiative started developing an "environmental contribution indicator" with the involvement of the ICT industry. The initiative was launched in 2007 with the aim to make positive changes in every aspect of production, society and national life through the application of ICTs.

The environmental contribution indicator is defined by the following formula:

(environmental contribution) =
(efficiency ratio) × (number of sales) × (contribution ratio)

The efficiency ratio is the amount of CO_2 emissions reduced by the products/activities in comparison with the amount of emissions without them. The contribution ratio is a ratio of the company's contribution to CO_2 reduction from those products/activities throughout their production and consumption, which is shared among suppliers, final product manufacturers, distributors and consumers. The company's net impact is calculated by discounting part of the CO_2 emissions caused by the company by the environmental contribution:

(Net impact) = (CO_2 emission) – (environmental contribution)

The development of this indicator is expected not only to encourage the ICT industry to consider more systemic innovation beyond immediate costs and benefits but also to facilitate consumers' choices of energy-efficient products and services through visualisation of the net impact. The initiative also proposes an incentive scheme in which the government and companies can purchase the credits of environmental impact reduction from consumers who buy energy-efficient products.

Source: Ministry of Economy, Trade and Industry, Japan (METI) (2008b) "Green IT Initiative as a Policy to Provide a Solution", presentation at the OECD Workshop on ICTs and Environmental Challenges, 22-23 May 2008, Copenhagen.

Eco-efficiency indicators would be more valuable if concept and methodology were unified since they can serve managerial and operational purposes at the same time. Composite indices can also ensure that corporate management commits to sustainable manufacturing if operational indicators are play a prominent role in the indexing process.

The further development and standardisation of environmental valuation techniques such as environmental accounting could also be valuable as this would help companies combine economic and environmental concerns and identify positive synergies. It would facilitate more rational and positive decision making regarding investments in sustainable manufacturing activities.

Life cycle thinking has helped companies to consider environmental effects beyond their factory gates, but to date no indicator set is applied by companies which takes into account system-level impacts beyond a single product life cycle. To encourage "system innovation" (see Chapter 1), a set of indicators is needed to identify system-wide impacts of new production processes and products/services. The development of an "environmental contribution indicator" by Japan's Green IT Initiative is an encouraging step in this direction (Box 3.10).

In general, most SMEs and suppliers lack incentives to use sustainability indicators and face capacity gaps, but the same is true for many larger companies. They all need to start by collecting data for a minimum set of individual indicators and then adopt more advanced indicators step by step. The Lowell Centre for Sustainable Production suggests that companies can start by monitoring compliance and gradually begin to address resource efficiency and more complex indicators that cover social effects as well as supply chain and life cycle considerations.

Notes

1. Principle 8 of the Rio Declaration adopted at the UNCED states: "To achieve sustainable development and a higher quality of life for all people, States should reduce and eliminate unsustainable patterns of production and consumption and promote appropriate demographic policies."
2. For example, Singh (2008) indicates that there are more than 600 initiatives on indicators and frameworks for the sustainable development of societies.
3. *Parameter*: A property that is measured or observed.

   Indicator: A parameter or value derived from parameters, which points to, provides information about, or describes the state of a phenomenon/ environment/area, with a significance extending beyond that directly associated with a parameter value.

   Index: A set of aggregated or weighted parameters or indicators.
4. A guide for constructing and using composite indicators for policy makers, academics, the media and other interested parties was prepared jointly by the OECD and the Joint Research Centre of the European Commission (OECD, 2003).
5. A number of Japanese electronics companies applying the "Factor" concept define eco-efficiency as a ratio of "product value" (*e.g.* functions in case of Panasonic) created per unit of environmental impact, instead of using economic value or cost as presented above (Shibaike *et al.*, 2008).
6. These concerns led to the launch of the Responsible Care initiative, first conceived in Canada in 1985 to address public concerns about the manufacturing, distribution and use of chemicals worldwide.

References

Ahmad, M.M. and N. Dhafr (2002), "Establishing and Improving Manufacturing Performance Measures", *Robotics and Computer Integrated Manufacturing*, Vol. 18, pp. 171–176.

Bringezu, S. (2003), "Industrial Ecology and Material Flow Analysis: Basic Concepts, Policy Relevance and Some Case Studies", in D. Bourg and S. Erkman (eds.), *Perspectives on Industrial Ecology*, pp. 20-34, Greenleaf Publishing, Sheffield.

Canon (2007), *Canon Sustainability Report 2007*, Canon Inc., Tokyo.

Celent (2007), *Socially Responsible Investing in the US and Europe: Same Goals but Different Paths*, Celent, Paris.

Dow Jones Indexes and Sustainable Asset Management (SAM) (2009), *Sustainability World Index Guide Book* (version 11.1), September, SAM, Zurich.

Environmental Protection Agency, United States (EPA) (1995), *An Introduction to Environmental Accounting as A Business Management Tool: Key Concepts and Terms,* EPA, Washington, DC.

European Aluminium Association (EAA) (2006), *Sustainability of the European Aluminium Industry 2006*, EAA, Brussels.

European Chemical Industry Council (CEFIC) (2007), *Responsible Care: Building Trust – Annual Report Europe 2006/2007*, CEFIC, Brussels.

Epstein, M.J. and M.J. Roy (2001), "Sustainability in Action: Identifying and Measuring the Key Performance Drivers", *Long Range Planning*, Vol. 34, pp. 585-604.

Figge, K. and T. Hahn (2004), "Sustainable Value Added: Measuring Corporate Contributions to Sustainability beyond Eco-Efficiency", *Ecological Economics*, Vol. 48, pp. 173-187.

Hamner, B. (2005), *Integrating Market-Based Sustainability Indicators and Performance Management Systems*, Cleaner Production International, Seattle, WA.

Hauschild, M. *et al.* (2005), "From Life Cycle Assessment to Sustainable Production: Status and Perspectives", *International Academy for Production Engineering (CIRP) Annals*, Vol. 54, No. 2, pp. 535-555.

Huppes, G. (2007), "Why We Need Better Eco-Efficiency Analysis: From Technological Optimism to Realism", *Technikfolgenabschätzung: Theorie und Praxis*, Vol. 16, No. 3, pp. 38-45.

International Federation of Accountants (IFAC) (2005), *Environmental Management Accounting*, IFCA, New York.

International Iron and Steel Institute (IISI) (2005), *Steel: The Foundation of a Sustainable Future – Sustainability Report of the World Steel Industry 2005*, IISI, Brussels.

International Organization for Standardization (ISO) (1997), *ISO 14040: Environmental Management – Life Cycle Assessment – Principles and Framework*, ISO, Geneva.

Jónsdóttir, H. *et al.* (2005), "Environmental Benchmarking a Tool for Continuous Environmental Improvements in the SME Sector", *NT technical report 588*, Nordic Innovation Centre, Oslo.

Krajnc, D. and P. Glavič (2005a), "A Model for Integrated Assessment of Sustainable Development", *Resources, Conservation and Recycling*, Vol. 43, pp. 189-208.

Krajnc, D. and P. Glavič (2005b), "How to Compare Companies on Relevant Dimensions of Sustainability", *Ecological Economics*, Vol. 55, pp. 551-563.

Kytzia, S. (2003), "Material Flow Analysis as a Tool for Sustainable Management of the Built Environment", in M. Koll-Schretzenmayr, M. Keiner and G. Nussbaumer (eds.), *The Real and the Virtual World of Spatial Planning,* Springer-Verlag, Berlin.

Matthews, H.S. and L.B. Lave (2003), "Using Input-Output Analysis for Corporate Benchmarking", *Benchmarking: An International Journal*, Vol. 10, No. 2, pp. 152-167.

McCool, S.F. and G.H. Stankey (2004), "Indicators of Sustainability: Challenges and Opportunities at the Interface of Science and Policy", *Environmental Management*, Vol. 33, No. 3, pp. 294-305.

Ministry of Economy, Trade and Industry, Japan (METI) (2008a), "The Response from the Ministry of Economy, Trade and Industry (METI) to the OECD Questionnaire Regarding Tools for Sustainable Manufacturing", METI, Tokyo.

METI (2008b) "Green IT Initiative as a Policy to Provide a Solution", presentation at the OECD Workshop on ICTs and Environmental Challenges, 22-23 May, Copenhagen.

Ministry of the Environment, Japan (MoE) (2005), *Environmental Accounting Guidelines 2005*, MoE, Tokyo.

OECD (2003), "Composite Indicators of Country Performance: A Critical Assessment", internal working document, OECD, Paris.

OECD (2005), *Environment at a Glance: OECD Environmental Indicators*, OECD, Paris.

OECD (2006), "Sustainable Production/Manufacturing Proposal to OECD", proposal submitted by the United States, internal working document for the Committee on Industry and Business Environment.

OECD (2007), "A Study on Methodologies Relevant to the OECD Approach on Sustainable Materials Management", OECD, Paris, *www.olis.oecd.org/olis/2007doc.nsf/linkto/env-epoc-wgwpr(2007)5-final*.

OECD (2008a), *Measuring Material Flows and Resource Productivity: Synthesis Report*, OECD, Paris.

OECD (2008b), *Measuring Material Flows and Resource Productivity: Volume I. The OECD Guide*, OECD, Paris.

Olsthoorn, X. *et al.* (2001), "Environmental Indicators for Business: A Review of the Literature and Standardisation Methods", *Journal of Cleaner Production*, Vol. 9, pp. 453-463.

Palme U. and A.M. Tillman (2008), "Sustainable Development Indicators: How Are They Used in Swedish Water Utilities?", *Journal of Cleaner Production*, Vol. 16, pp. 1346-1357.

Peele, C. (2005), "Substance Flow Analysis and Material Flow Accounts", Background paper for the Framing a Future Chemicals Policy Workshop, Boston.

Putnam, D. and A. Keen (2002), "ISO 14031: Environmental Performance Evaluation", draft submitted to *Journal of the Confederation of Indian Industry*, Altech Environmental Consulting, Toronto.

Rao, P. *et al.* (2006), "Environmental Indicators for Small and Medium Enterprises in the Philippines: An Empirical Research", *Journal of Cleaner Production*, Vol. 14, pp. 505-15.

Richardson, B.J. (2008), *Socially Responsible Investment Law: Regulating the Unseen Polluters*, Oxford University Press, Oxford.

Schaltegger, S. and M. Wagner (2006), *Managing and Measuring the Business Case for Sustainability: Capturing the Relationship between Sustainability Performance, Business Competitiveness and Economic Performance*, Greenleaf Publishing, Sheffield.

Schmidheiny, S. (1992), *Changing Course*, MIT Press, Cambridge, MA.

Schmidt, W.P. (2008), "Developing a Product Sustainability Index", in OECD (ed.), *Measuring Sustainable Production*, OECD Sustainable Development Studies, pp. 115-126, OECD, Paris.

Schwarz, J. *et al.* (2002), "Use Sustainability Metrics to Guide Decision-making", *Chemical Engineering Progress*, American Institute of Chemical Engineers.

Shibaike, N. *et al.* (2008), "Activity of Japanese Electronics Industry on Environmental Performance Indicator toward Future Standardization", Proceedings of Electronics Goes Green 2008+, pp. 473-477, Fraunhofer IZM, Berlin.

Singh, R.K. *et al.* (2007), "Development of Composite Sustainability Performance Index for Steel Industry", *Ecological Indicators*, Vol. 7, pp. 565-588.

Singh, R.K. (2008), "Developing a Composite Sustainability Index", in OECD (ed.), *Measuring Sustainable Production*, pp. 97-114.

Social Investment Forum (2008), 2007 *Report on Socially Responsible Investing Trends in the United States: Executive summary*, Social Investment Forum, Washington, DC.

The Climate Group (2008) *SMART 2020: Enabling the Low Carbon Economy in the Information Age*, report by the Climate Group on behalf of the Global e-Sustainability Initiative (GeSI), The Climate Group, London.

United Nations Division for Sustainable Development (UNDSD) (2001), *Environmental Management Accounting Procedures and Principles*, United Nations, New York.

United Nations, European Commission, International Monetary Fund, OECD and World Bank (2003), *Integrated Environmental and Economic Accounting 2003,* United Nations, New York.

Veleva, V. and M. Ellenbecker (2001), "Indicators of Sustainable Production: Framework and Methodology", *Journal of Cleaner Production*, Vol. 9, pp. 519-549.

Wackernagel, M. (2008), "Measuring Ecological Footprints", in OECD (ed.), *Measuring Sustainable Production*, pp. 49-59.

World Business Council for Sustainable Development (WBCSD) (2000), *Eco-efficiency: Creating More Value with Less Impact*, WBCSD, Geneva.

Chapter 4

Measuring Eco-innovation: Existing Methods for Macro-level Analysis

Quantitative measurement can be very important for understanding the complex and diverse nature of eco-innovation. This chapter reviews existing methods for measuring eco-innovation at the macro level and analyses their strengths and weaknesses. Because capturing overall patterns of eco-innovation raises significant challenges, it is important to apply different analytical methods, possibly combined, and view information from various sources (generic data and specially designed surveys), taking careful account of the context of the data.

Introduction

Eco-innovation is a new concept of great importance for both industry and policy makers. It offers them a means of moving industrial production in a more sustainable direction and systematically responding to global environmental challenges such as climate change. As defined in Chapter 1, eco-innovation can concern all types of innovations that lower environmental impact as compared to relevant alternatives. Such innovations may be technological or non-technological (marketing, organisational or institutional) and can be motivated by economic or environmental considerations, or both.

Quantitative measures of an activity are an important input for informed decision making by policy makers and other stakeholders. Quantitative analysis is increasingly used to understand general innovation activities (*e.g.* OECD, 2008a; EC, 2008) and would also be important for understanding eco-innovation. This chapter therefore reviews existing quantitative methods for measuring eco-innovation at the macro level (*i.e.* sectoral, local and national). It also examines the strengths and weaknesses of existing methodologies and offers future directions for improving the measurement of eco-innovation.

This chapter starts by briefly outlining the reasons for and benefits of measuring eco-innovation (*why measure?*). Second, it introduces various aspects of eco-innovation that can be measured quantitatively (*what to measure*). Third, it presents four major ways to capture eco-innovation through existing data sets and statistics (*how to measure*) with examples of such measurements and their strengths and weaknesses. Fourth, the use of surveys is considered as an alternative means of obtaining data on eco-innovation, and existing surveys on eco-innovation, such as the new "eco-innovation module" added in the European Union's (EU) Community Innovation Survey 2008, are reviewed. A brief conclusion follows.[1]

Benefits of measuring eco-innovation

Faced with rising costs for using natural resources and managing emissions and wastes, the competitiveness of firms, regions and countries is increasingly linked to their ability to drive eco-innovation. Yet, environmental technologies have been largely neglected in economic statistics, and very little is therefore known about growing world trade in environmentally beneficial goods and services. Nor is much known about the adoption of innovations to reduce the environmental impacts of firms, sectors and countries or the environmental improvements achieved thanks to the creation and application of eco-innovations.

Measurement could help evaluate progress in various categories of eco-innovation – for example, to assess which countries are leaders in promoting eco-innovation, or how much progress countries are making to decouple economic growth from environmental degradation. It may also allow for an analysis of drivers of eco-innovation, including environmental legislation and regulations, and of the economic consequences. Measuring eco-innovation can:

- help policy makers understand, analyse and benchmark overall trends in eco-innovation activities (*e.g.* increasing, decreasing, transitions from end-of-pipe towards cleaner production, changes in business models), as well as trends in specific product categories (*e.g.* wind turbines).
- help policy makers identify drivers of and barriers to eco-innovation. This information can inform the design of effective policies and framework conditions.
- raise awareness of eco-innovation among businessmen, policy makers and other stakeholders and encourage companies to increase eco-innovation efforts on the basis of an analysis of its benefits.
- help society to tackle global environmental challenges by making the environmental improvement that has been or can be achieved through eco-innovation more tangible to producers and consumers alike.

Aspects of eco-innovation to measure

Eco-innovation includes both environmentally motivated innovations and unintended environmental innovations. The environmental benefits of an innovation may thus be a side effect of other goals such as reducing costs for production or waste management. Eco-innovations may also arise from institutional changes in values, knowledge, norms and administrative actions or from new stakeholder collaborations. In fact, almost all firms can become eco-innovators.

This broad definition of eco-innovation may create a problem for analysts who prefer definitions that are limited to a single type of activities. The definition of innovation in the *Oslo Manual* (OECD and Eurostat, 2005) has similarly been criticised for defining innovation so broadly that almost all firms could be innovators. For example, that definition ranges from new off-the-shelf technology purchased by a firm for the first time to long-term research and development (R&D) projects; it also includes both technological and non-technological innovation.

Part of the definitional problem arises because innovation is relative. The first-time use of a pollution control device by a firm is an innovation from the viewpoint of that firm, but not of the manufacturer of the device. For the manufacturer, what counts as an innovation is a significant change in the pollution control device or the creation of a new technology. When measuring eco-innovation, it should be made clear whether one is measuring the creation of an innovation or the first implementation of products, technologies, services or practices. Another important distinction is whether the innovation is an incremental improvement of something that already exists or is entirely new.

This definitional problem would be solved by collecting sufficient data to be able to identify:

- how firms eco-innovate, or the nature of eco-innovation;
- the drivers and barriers that affect different types of eco-innovations;
- the impacts of different types of eco-innovations.

The following sections explain each of these three aspects in detail (Figure 4.1).

Figure 4.1. Aspects of eco-innovation to measure

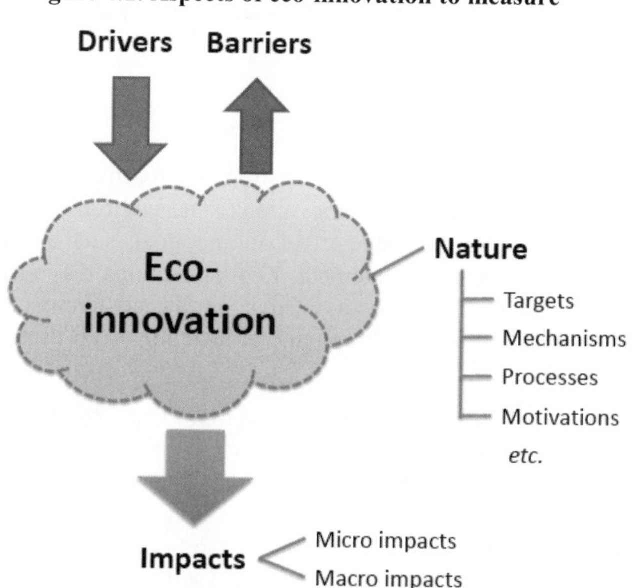

Nature of eco-innovation

Each eco-innovation is unique in some sense. Different attempts have been made to analyse the diverse nature of eco-innovation by constructing a classification of eco-innovations. Based on the *Oslo Manual*, Chapter 1 categorises eco-innovations according to the "targets" for innovation into product, process, marketing, organisational and institutional innovation. It also introduces another axis of categorisation, the "mechanisms" used by firms to introduce eco-innovations either by modifying existing technology (incremental innovation) or by creating entirely new solutions or even changing business models (radical innovation). As combinations of such targets and mechanisms, Chapter 1 also classifies different types of eco-innovation processes in the manufacturing sector, from pollution control through cleaner production and life cycle thinking to closed-loop production and industrial ecology. Another distinction is whether eco-innovations are environmentally motivated or initiated for non-environmental reasons.

The European Commission (EC)-funded Measuring Eco-Innovation (MEI) project created a classification according to the purposes or objectives of eco-innovations. It distinguishes between *i)* environmental technologies; *ii)* organisational innovations for the environment; *iii)* product and service innovations that offer environmental benefits; and *iv)* green system innovation. The first three can be measured in principle and thus inform policy makers about changes in the nature of eco-innovation, such as a shift from curative (end-of-pipe) solutions to preventive (cleaner production) solutions. Green system innovations are the most difficult to measure as they are not about identifiable innovations but about evolving systems that entail multiple changes.

It is also possible to categorise some types of eco-innovations as "environmental goods". However, it is difficult to reach broad agreement on the definition of environmental goods, mainly because many candidate goods have a range of uses besides environmental protection. More significantly, environmental goods are often designated as such in relation to a conventional alternative that may well be included in the very same classification. The OECD (2008b) therefore argues that commodity classifications cannot be used to develop indicators for measuring eco-innovation.

Another simple system focuses on the processes of innovations and divides them into end-of-pipe and cleaner production innovations (Frondel *et al.*, 2004). Results from a 2003 OECD survey in seven industrialised countries (Canada, France, Germany, Hungary, Japan, Norway and the United States) found that cleaner production technology accounted for between 58% (Germany) and 87% (Japan) of the total number of process innovations with environmental benefits (Figure 4.2). In Germany, investment in end-of-pipe technology has fallen. This appears to be partly because of an increase in investment in cleaner production technology.

Figure 4.2. Types of environmental technologies implemented in seven OECD countries

Percentages

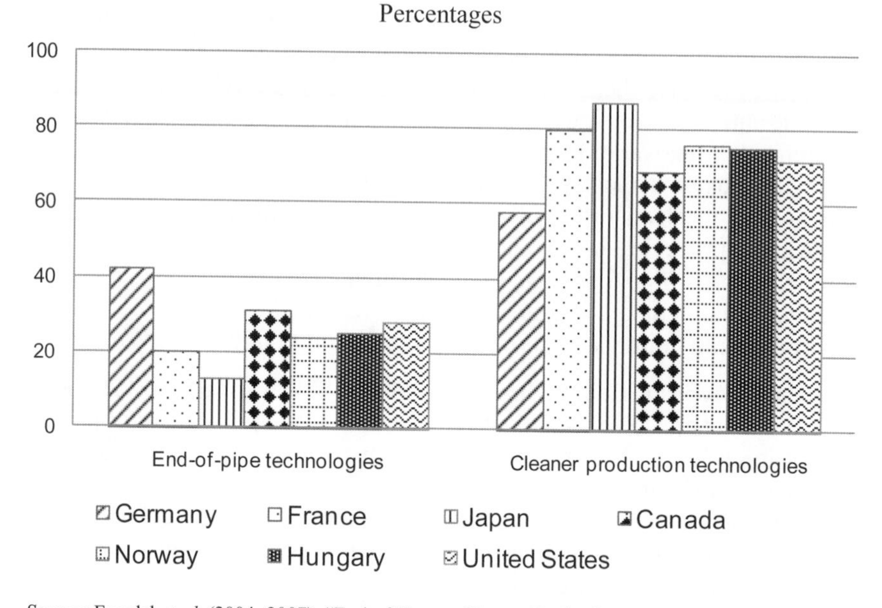

Source: Frondel *et al.* (2004, 2007), "End-of-Pipe or Cleaner Production? An Empirical Comparison of Environmental Innovation Decisions across OECD Countries", *Business Strategy and the Environment*, Vol. 16, No. 8, based on data from a total of 3 100 establishments.

Past research and measurement activities have mostly focused on pollution control and abatement. Eco-innovation research and data collection should not, however, be limited to products from the environmental goods and services sector or to environmentally motivated innovations, but should cover all innovations with environmental benefit. Research should inquire into the nature of the benefits and the motivations. The many types of eco-innovations also require a variety of indicators to obtain a full picture of the eco-innovative efforts of firms. An indicator that only covers end-of-pipe innovations, for example, would fail to miss the apparent shift in Germany towards integrated cleaner production.

Drivers of and barriers to eco-innovation

Rennings and Zwick (2003) define five drivers of eco-innovation: regulation, demand from users, capturing new markets, cost reduction and image. Determinants for different kinds of eco-innovation were also studied in the EC-funded IMPRESS project.[2] This survey found that many reasons for introducing eco-innovation besides complying with regulations. They include: improving the firm's image; reducing costs; achieving accreditation; as part of product and service innovations; securing existing markets; and increasing market share. Compliance with environmental regulations was more important for pollution control innovations than for other types of eco-innovation, especially service, distribution and product innovations.

On the other hand, the EU's Environmental Technologies Action Plan (ETAP) refers to the following barriers to the introduction and dissemination of environmental technologies (EC, 2004):

- **Economic barriers,** ranging from market prices which do not reflect the external costs of products or services (such as health-care costs due to urban air pollution) to the higher cost of investments in environmental technologies because of their perceived risk, the size of the initial investment, or the complexity of switching from traditional to environmental technologies.
- **Regulations and standards** may act as barriers to innovation when they are unclear or too detailed, while good legislation can stimulate environmental technologies.
- **Insufficient research effort,** coupled with inappropriate functioning of the research system and weaknesses in information and training.
- **Inadequate availability of risk capital** to move from the drawing board to the production line.
- **Lack of market demand** from the public sector as well as from consumers.

Ashford (1993) provides a more comprehensive list of barriers than that of the ETAP. As such barriers tend to be interrelated, it is not necessarily easy for policy makers and industry to tackle them. They include:

- **Technological barriers** such as a lack of available technology or performance capabilities;
- **Financial barriers** such as high costs of research, inability to predict future liability costs, impact on competitiveness, or a lack of economies of scale;
- **Labour force-related barriers** such as a lack of knowledgeable management or reluctance to employ trained engineers;
- **Regulatory barriers** such as disincentives to invest in recycling, regulatory uncertainty, focus on end-of-pipe treatments;

- **Consumer-related barriers** such as tight product specifications or risk of losing customers owing to a change in product characteristics;
- **Supplier-related barriers** such as a lack of support for maintenance;
- **Managerial barriers** such as a lack of co-operation among different functions within the firm, a reluctance to change operating methods, or a lack of education and training of employees.

Impacts of eco-innovation

Eco-innovation should help decouple economic growth from environmental degradation and create win-win solutions. The identification of the impacts of eco-innovation on economic growth and employment is, however, not straightforward and is likely to vary, depending on the types of eco-innovations and the context in which they are used. Eco-innovation may create more jobs and economic wealth in the producing sector, but if the innovation increases costs for users, the eco-innovation may not be sufficient to compensate for losses elsewhere. For example, Germany has a flourishing solar and wind power industry thanks to the renewable energy feed-in law which establishes high prices for green electricity fed into the grid, but as a result German consumers and industry pay higher prices for electricity than they otherwise would. More expensive electricity might hamper the competitiveness of other sectors that are intensive users of electricity.

Nor is the identification of the environmental impacts of eco-innovation always easy. It is important to recall that, so far, many eco-innovations may have helped to achieve a *relative decoupling* in OECD countries, with emissions levels falling relative to economic growth, but impacts have been increasing in absolute terms in most countries for many pollutants. Achieving *absolute decoupling* requires not only reductions achieved by eco-innovation at the micro level but also averting "rebound effects" at the macro level.

Whereas companies are mostly interested in impacts at the micro level, policy makers are generally more interested in macro-level impacts. The links between micro and macro impacts are complex, with many cross-sectoral impacts and feedback loops such as:

- Cost-saving eco-innovations generate wealth that will be spent on goods and services that can have a negative environmental impact, creating second-order environmental burdens.
- Cost-increasing eco-innovations are likely to contribute more to absolute decoupling but possibly at the expense of lower economic growth.

- Even though many new products are more environmentally benign than old ones, overall environmental gains will be counterbalanced by the economic growth arising from those innovations.
- Full accounting of the impacts of eco-innovation requires life cycle analysis along the entire value chain, from resource extraction to waste management.
- Micro-level behaviour can be affected by macro-level factors such as taxes and regulations.

Use of generic data sources to measure eco-innovation

Eco-innovation can be measured and analysed by utilising the following four categories of data. They are based on the "input" to and the "output" from eco-innovations and the "impact" of eco-innovation:

- **input measures,** *e.g.* R&D expenditures, R&D personnel, other innovation expenditures (such as investment in intangibles including design expenditures and software and marketing costs);
- **intermediate output measures,** *e.g.* the number of patents or numbers and types of scientific publications;
- **direct output measures**, *e.g.* the number of innovations, descriptions of individual innovations, sales of new products from innovations;
- **indirect impact measures**, *e.g.* changes in resource efficiency and productivity.

These data can be obtained by using widely available generic sources of data which are not collected specifically to measure eco-innovation and by conducting surveys specifically designed to measure eco-innovation (Figure 4.3). This section reviews methodologies for using generic data sources, and the next section reviews survey methodologies. Each methodology is explained with examples of existing research.

Figure 4.3. Options for measuring eco-innovation

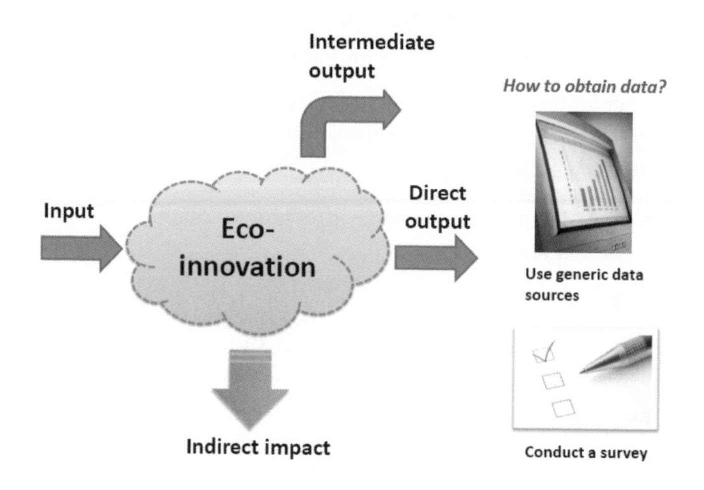

Input measures

R&D statistics are widely used in innovation research, but they have a few limitations. They tend to capture formal R&D activities typically carried out in formal laboratories in manufacturing companies and to underestimate R&D activities conducted by smaller firms or in the services sector, which are often implemented on a more informal basis (Kleinknecht *et al.*, 2002). Furthermore, R&D data do not cover non-technological innovation activities such as marketing and organisational and institutional eco-innovations.

Data for "environmental R&D" are very limited in scope. The only consistent data across OECD countries are those for government budget appropriations or outlays for R&D (GBAORD) under "control and care for the environment". These refer to budget provisions instead of actual expenditure. The data include both current and capital expenditure and cover not only government-financed R&D performed in government establishments, but also government-financed R&D in the business enterprise, non-profit and higher education sectors, as well as abroad (Wilén, 2008).

For the private sector, environmental R&D can be defined in two ways: R&D that is environmentally motivated and R&D that is relevant for reducing environmental impact either in the company or elsewhere (*e.g.* at the point of use). Both types of statistics would be of value but neither is available on a consistent basis from generic data sources such as official R&D surveys. Research from specialised surveys suggests that official R&D surveys could collect some types of data on environmental R&D by the private sector (see the following section).

Intermediate output measures

Intermediate output measures consist of patents and scientific publications and citations. Patent data are the most commonly used to construct intermediate indicators for *inventions* (Dodgson and Hinze, 2000). A patent is an exclusive right to exploit (make, use, sell or import) an invention over a limited period of time (20 years from filing) in the country in which the application is made. Patents are granted for inventions which are novel, inventive and have an industrial application (OECD, 2004) but they need not be commercially applied. Consequently, they are not direct measures of innovations. Furthermore, the standard of novelty and utility for granting a patent is not necessarily high. The European Patent Office (EPO) grants patents for about 70% of the total applications, while the US Patent and Trademark Office (USPTO) grants patents for about 80% of patent applications.

On the other hand, patents have several advantages over R&D expenditures: *i*) they explicitly give an indication of inventive output; *ii*) they can be disaggregated by technology group; and *iii*) they combine detail and coverage of technologies (Lanjouw and Mody, 1996). Moreover, they are based on an objective and slowly changing standard because they are granted on the basis of novelty and utility (Griliches, 1990).

Patent counts can therefore be used as an indicator of the level of innovative activity in the environmental domain. As for innovation in general, patents covering *eco-inventions* can be used to measure research and inventive activity and to study the direction of research in a given technological field. Whether or not something is an eco-innovation depends on its environmental impacts; therefore, to be recognised as an "eco-patent", the environmental gain must be described or there must be pre-existing data on the environmental benefits of a patent class. Otherwise, inventions with non-intentional environmental benefits will not be identified in patent analysis.

The MEI project proposes the following four-step method for screening "eco-patents" (MERIT *et al.*, 2008):

1. Choose relevant parameters (*e.g.* a pollutant such as sulphur dioxide [SO_2] or an environmental technology such as wind power).
2. Search patents using keywords based on relevant environmental technology aspects in order to generate a set of potentially relevant patents.
3. Screen the abstracts of the patents generated to determine whether they are relevant and exclude irrelevant patents.

4. Retrieve patent families. These are patent applications filed in countries other than the home country. This helps to exclude patents of minor importance.

Similar methods can be applied to scientific publications of firms. These can signal scientific competence and/or interest in scientific communication in a specific area. Collaboration between scientific and industrial institutions can be measured by co-publication of publications or patents (Dodgson and Hinze, 2000).

The OECD has been active in the creation of eco-innovation statistics based on patent analysis. International Patent Classification (IPC) classes have been identified for selected environmental technologies: alternative vehicle propulsion, climate change mitigation technologies and a wide range of other environmental technologies. Whereas past research focused on pollution control technologies, recent research focuses on renewable energy technologies and alternative fuel vehicle (AFV) technologies.

An important new development is the creation of the EPO/OECD Patent Statistical Database (PATSTAT) which contains 70 million patent applications from 80 countries. This database can be used to identify both end-of-pipe environmental inventions and "more integrated technological innovations" with environmental benefits such as fuel cells for motor vehicles (OECD, 2008b).

Patent analyses can also be used for measuring technology transfer. The idea of using patent data to measure international technology transfers arises from the fact that there will be a partial "trace" of the three identified channels of technology transfer (trade, foreign direct investment and licensing) in patent applications. OECD (2008b) proposes to use "duplicate patents" (obtained in several countries) as a measure for technology transfer. There is a positive correlation between duplicate patents and exports of wind power technologies.

There are a number of limitations on the use of patent data. Not all eco-innovations can be identified through patents. Environmental patents mainly measure *inventions* that underlie some, but not all, green product innovations and end-of-pipe technologies. However, for organisational and process innovations, patent analysis is much less useful, as many of these innovations are not patented.

Furthermore, the potential commercial value of patents varies substantially. Different methods can be used to assess a patent's value. For example, one can ask patent owners about past returns and the potential market value of their rights, look at patent renewals, or use the number of citations as a proxy for commercial value. Here, the development of the

OECD Triadic Patent Family Database is of great interest since it provides a database of "quality" inventions. The use of patent families – *i.e.* patent applications with the same priority date filed in different countries – makes it possible to focus on the most valuable innovations. Because of the added costs of filing abroad, less valuable patents are usually filed only in the inventor's own country.

Direct output measures

Direct output measures cover the content and scale of actual eco-innovations. Announcements in trade journals[3] and product information databases are important generic sources of information on the content and scale of eco-innovations. An example is Yahoo!'s green car database.

Very few product databases contain environmental information. For specific types of products, a database of eco-innovation output could be created by sampling the new product announcement sections of technical and trade journals or by examining product information provided by producers. The strengths of the product announcement sampling method are:

- It measures actual innovations introduced into the market place.
- The indicator is timely: the timing of announcements is close to the date of commercialisation.
- The data are relatively cheap to collect and do not require direct contact with the innovative firms.
- From the description, it is possible to infer information about the innovation, such as whether it is an incremental or radical innovation, and what the performance characteristics are.

There are also some limitations:

- The existence of an adequate selection of journals is necessary to ensure comprehensive coverage.
- In-house process innovations are rarely reflected in technical and trade journals.
- Although the number of innovations can be counted, appreciation of their importance is subjective.

Information from trade journals is often available in electronic form. Information about products may also be available on the Internet. This can allow researchers to track the evolution of products' performance characteristics. Digital announcements and consumer information databases are a neglected source of innovation output indicators that could be more intensively exploited to produce useful metrics.

It is important to note that these generic sources rarely, if ever, provide output measures in terms of revenue or the effect of eco-innovation on production costs. Such measures require specialised surveys (discussed in the following section).

Indirect impact measures

Eco-innovation can be indirectly measured on the basis of data on changes in absolute environmental impact or in resource productivity. Eco-efficiency is one of the most popular ways to capture resource productivity and is usually measured at the product or service level (see Chapter 3). A common definition of eco-efficiency is: "less environmental impact per unit of product or service value" as indicated below (WBCSD, 2000).

$$\textit{Eco-efficiency} = \frac{\textit{environmental impact}}{\textit{product or service value}}$$

An improvement in the eco-efficiency ratio is indicative of eco-innovation. Such ratios can be determined for company processes, products, sectors and nations. The ratio can be calculated from generic data at the sectoral level or national level, using data for value added and emissions from national accounting systems as well as specialised survey data. It may also be increasingly possible to construct performance benchmarks for individual firms, using microdata from their sustainability reports. A challenge for benchmarking based on microdata is to cover environmental aspects over the entire value chain as this requires combining data from different companies. To be meaningful for benchmarking, data from single companies have to be broken down for functional units (a product or production process).

Instead of eco-efficiency indicators, similar indicator methodologies can be used to monitor resource productivity, including ecological footprint, material flow analysis (MFA), material input per service unit (MIPS) and ecological rucksack (Mill and Gee, 1999; see Chapter 3). It is important to note, however, that there is no simple causal relation between eco-innovations and eco-efficiency, as changes in eco-efficiency may reflect factors such as sectoral changes and non-innovative price-based substitution.

Overall evaluation and suggestions

Generic data sources are best suited for providing data on certain aspects of eco-innovation, such as investment in types of eco-innovations, or on the number of different types of intermediate and marketed eco-innovations. In contrast, none of the generic data sources provides information on the drivers of and barriers to eco-innovation, on revenue, or on the effect of eco-innovation on production costs, and only a few provide information on the impacts of eco-innovation. Although some methods may be better than others, no single indicator derived from generic data sources is an ideal measure of eco-innovation as each has its strengths and weaknesses. To understand overall patterns of eco-innovation and the drivers of those patterns, it is important to view different indicators together, possibly by mapping data, listing headline indicators or developing a composite index.

More effort could be devoted to obtaining direct measures of eco-innovation outputs using generic documentary and digital sources in addition to those for innovation inputs (such as R&D expenditures) or intermediary outputs (such as patent grants). Eco-innovation can also be monitored indirectly by changes in resource efficiency and productivity. These two avenues have been underexplored and could be used to augment the current rather narrow knowledge base.

Methods for measuring eco-innovation should be combined. Concrete suggestions for combining measures and methods are:

- Contact a sample of inventors and ask questions about their patents, such as the extent to which their efforts are spurred by specific regulations, environmental concerns, their economic gain, etc.
- Compare patent patterns with R&D patterns and data about innovation output collected through analysis of documentary and digital sources. This would help assess the value of patent analysis and obtain more robust research findings based on multiple data sources.
- Combine macro-level information on eco-efficiency with microdata from companies about technological and non-technological eco-innovation to better understand the links between micro and macro measures.
- Combine information on general innovation investments with information on eco-innovation and environmental performance.

The ability to link data from different databases could substantially improve studies on eco-innovation. For example, OECD (2008b) suggests that it should be possible to link firms in the PATSTAT database to other datasets that contain information on each firm's employment levels and profitability. This would allow for an analysis of the impact of eco-innovation (proxied by patents) on firm performance.

Use of surveys to measure eco-innovation

Unlike existing data and statistics, surveys on the eco-innovation activities of firms may provide researchers with more detailed information on a number of aspects of eco-innovation, such as investment in different types of eco-innovation and information on drivers, barriers and impacts of eco-innovation. These data would permit econometric analysis of the effect of different drivers on outcomes. Survey results at the level of the enterprise or establishment can also be aggregated to provide sectoral, regional or national statistics.

This section reviews the different approaches taken in past surveys of eco-innovation and evaluates their strengths and weaknesses. It introduces the next EU Community Innovation Survey (CIS), which includes an optional one-page set of questions on eco-innovation and reviews national surveys of pollution abatement and control expenditures (PACE). It concludes by outlining the types of survey questions that could be introduced in the future.

Existing surveys on eco-innovation

There are two basic sources of survey indicators.[4] The first consists of official, large-scale innovation surveys that sample thousands of firms and are performed on a regular basis. The second consists of smaller one-off surveys by academics, research institutes or government agencies. These usually focus on a limited geographical region or set of sectors.

Large-scale national innovation surveys in Europe and in Australia, Canada, Japan, Korea and New Zealand include a few questions that are relevant to eco-innovation. For example, the EU's 2006 CIS asked about the importance of the "effects of your product and process innovations" to "reduce materials and energy per unit output" and to "reduce environmental impacts or improve health and safety". Unlike the PACE data (discussed below) and many patent analyses, these questions provide information on the prevalence of innovation with environmental benefits without limiting the results to *intentional* eco-innovation. Furthermore, the information on eco-innovation can be linked to other firm-level innovation strategies and characteristics. The main disadvantage of these surveys is that, so far, they have only collected data on reductions in material and energy use or "reduced environmental impacts" in general. Moreover, the last question unfortunately combines environmental impact with a possibly unrelated effect on health or safety.

Several past smaller surveys, summarised in Table 4.1, have examined eco-innovation in far greater depth.[5] Most have not queried firms about their in-house innovative activities but have covered the adoption of environmental technologies for internal process improvement (pollution control technologies or cleaner processes). For each survey, the table describes the target population of firms, the number of responses and the response rate, and the types of questions asked. For example, it notes if the survey included questions about the type of innovation (management system, adoption of technology, technology developed in house), the motivations for or drivers of eco-innovation, the economic effects of eco-innovation, and the source of knowledge or barriers to eco-innovation. As the third column shows, many specialised environmental surveys cannot match the response rates of official innovation surveys. Low response rates reduce confidence in the accuracy of prevalence rates. One way to address this problem is to conduct a non-response analysis to determine if non-respondents differ in any significant way from respondents. To date, this technique has rarely been used in eco-innovation surveys.

Among the surveys listed in Table 4.1, four focus specifically on eco-innovation (Green *et al.*, 1994; Lefebvre *et al.*, 2003; Rennings and Zwick, 2003; Johnstone, 2007). The fifth covers biotechnology in general but asks a large number of questions on eco-innovation (Arundel and Rose, 1999). These are the only five studies that differentiate between innovation as creation and as adoption.

Most of these small surveys focus on the motivation for and drivers of eco-innovation, followed by its impact on costs, employment or skills. All three studies on employment and skills (Pfeiffer and Rennings, 2001; Getzner, 2002; Rennings and Zwick, 2003) concern Europe. None obtains interval-level data on employment effects (such as percentage changes in job gains or losses) because respondents can rarely provide accurate estimates. Instead, the survey questions ask for data either by category (employment increased or decreased with percentage categories such as between 10% and 25%) or by nominal level data (employment increased or decreased, yes or no). As an example, Pfeiffer and Rennings (2001) report that between 84% and 91% of German firms (depending on the type of eco-innovation) found that the innovation had no effect on employment; less than 5% reported a decrease.

Table 4.1. List of existing eco-innovation surveys

Reference	Target firms	Responses (response rate)	Type of innovation	Motivations and drivers	Economic effects	Knowledge sourcing/ impediments
Steger, 1993	German manufacturing and service firms	592 (not given)	A	✓	C	
Green *et al.*, 1994	UK firms interested in government support programmes	169 (21%)	A, CR	✓		
Arundel and Rose, 1999	Canadian firms in sectors with potential biotechnology applications	2 010 (86%)	A, CR	✓	C	K, I
Blum-Kusterer and Hussain, 2001	German and UK pharmaceutical firms	32 (21%)	M	✓		I
Pfeiffer and Rennings, 2001	German manufacturing firms	400 (45%)	A	✓	E, S	
Getzner, 2002	EMAS/ISO firms in Austria, France, Germany, the Netherlands, Spain, Sweden	407 (16%)	A	✓	E, S	
Andrews *et al.*, 2002	SMEs in Australia	145 (29%)	M, A		C	K
Lefebvre *et al.*, 2003	SMEs in four industries in Canada	368 (quota sampling)	M, A, CR	✓		
Rennings and Zwick, 2003	Manufacturing and service firms in Germany, Italy, the Netherlands, Switzerland and the United Kingdom	1 594 (not given for all countries)	A, CR	✓	C, E, S	
Scott, 2003	US manufacturing firms	132 (16%)	RD	✓		K
Zutshi and Sohal, 2004	ISO 14001 firms in Australia and New Zealand	143 (46%)	M	✓		K, I
Johnstone, 2007; Frondel *et al.*, 2004	Companies in all manufacturing sectors with more than 50 employees	4 200 (25%)	M, A, CR, RD	✓	C	

Type of innovation: M: management systems, **A**: technology adoption, **CR**: technology creation (innovation developed in firm), **RD**: environmental R&D.
Economic effects: C: costs, **E**: employment, **S**: skills. **Motivation and drivers: ✓** = these elements were examined.
Knowledge sourcing/impediments: K: knowledge sourcing, **I**: impediments to adoption.
Source: Arundel *et al.* (2007), "Indicators for Environmental Innovation: What and How to Measure", in Marinova, Annandale and Phillimore (eds.), *International Handbook on Environment and Technology Management*, Edward Elgar, Cheltenham, updated with reference to Johnstone (2007) and Frondel *et al.* (2004).

Knowledge sourcing and impediments to eco-innovation have received the least attention in eco-innovation surveys. One exception is the survey by Andrews *et al.* (2002) which asked if firms shared their knowledge of and experience with cleaner production with other firms and with industry associations. This is a valuable area for future research if combined with data on licensing behaviour, because the policy goal of encouraging knowledge sourcing may conflict with a firm's strategic interest in keeping its eco-innovations secret.

The Statistics Canada survey (Arundel and Rose, 1999) on biotechnology applications is the only study to cover all three aspects of measuring eco-innovation. The respondents were asked if their firm currently used or planned to use one of five carefully defined environmental biotechnologies. Users of one or more of these technologies were then asked a series of questions on investment, their motivations for adopting the technology, difficulties with implementation, results from their use, and the principal internal and external sources of information to facilitate the adoption of environmental biotechnologies (Arundel and Rose, 1999).

The two largest specialised surveys on eco-innovation to date are that of the EC-funded IMPRESS project (mentioned above) and the OECD survey on environmental policy and firm-level management (Johnstone, 2007). The IMPRESS project conducted 1 594 telephone interviews with randomly selected industry and service firms in eight sectors from five European countries (Germany, Italy, the Netherlands, Switzerland and the United Kingdom). It obtained measures of the economic effects of "the most important environmental innovation" introduced by the company in the last three years by asking about the effect of the innovation on sales, prices and costs for energy, materials, waste disposal and labour. For example, the questions asked if the innovation increased (or decreased) sales by up to 5%, 5% to 25%, or by over 25%. The analysis identified both positive and negative economic effects of eco-innovation. The number of companies experiencing positive employment and economic effects was higher than the number of those experiencing negative effects (Rennings and Zwick, 2003).

The OECD survey covered links between governments' environmental policies and environmental management, investments, innovation and performance in private firms in manufacturing sectors in seven OECD countries (Canada, France, Germany, Hungary, Japan, Norway, and the United States). It used several criteria for identifying such links, including perceived stringency of the policy framework, number of inspections in the last three years, and the reported presence of targeted measures to encourage the use of environmental management systems or tools. This is also one of the few studies to have specifically looked into environmental R&D (Johnstone, 2007).

This survey asked firms about the share of their R&D budget spent on environmental conservation. Overall, 9% of facilities in the OECD study reported positive investments in environment-related R&D (Johnstone, 2007). It also obtained information on the amount of R&D expenditures for environmental purposes. In Japan, environment-related R&D expenditures accounted for 17% of total R&D expenditures in the manufacturing sector. The researchers compared this figure with the results from a Japanese R&D survey and found that the figures from the specialised survey were much higher: 17% vs. 3% (Arimura *et al.*, 2007). While specialised surveys may elicit more accurate responses than general surveys, they may also be subject to a substantial bias.

Since the term "environment" may be too general, the accuracy of responses to questions on environmental R&D might be improved by using specific categories such as waste reduction, efficiency in material use, and pollution prevention and control. In a survey of US manufacturing firms, Scott (2003) asked a series of questions on different types of environmental R&D aimed at reducing toxic air emissions. However, the low response rate (16%) suggested that the survey method was inappropriate either because it was too complex and did not match the accounting systems that firms use to manage their R&D investments, or because firms that have effectively integrated eco-innovation into their mainstream innovations had difficulty separating environmental R&D from other types of R&D.

An interesting avenue for future research on eco-innovation is to develop panel surveys that gather information from the same firms over time. A good example is the Mannheim Innovation Panel led by the ZEW which includes more than 1 800 Germany-based firms with at least some new product development activities. This is a bi-annual survey that provides important information about the introduction of new products, services and processes, expenditures for innovations, and how economic success is achieved with new products, new services and improved processes. In addition, the survey gives information about factors that promote and hinder innovation activities of enterprises (Horbach, 2008). The results of such surveys can permit sophisticated analysis of the effect of motivations and management systems on different types of eco-innovation.

CIS 2008 eco-innovation module

The EU's next CIS 2008, which covers innovation activities between 2006 and 2008, includes a new "eco-innovation module" (presented in Box 4.1). The module was developed in collaboration between the CIS Task Force of Eurostat, the EC's DG Environment, several academics in the MEI project and the UNU-MERIT.

The first question asks respondents if they have introduced an innovation with one or more environmental benefits. Six of these environmental benefits are achieved during the use of the innovation by the enterprise and three during the use of the innovation by the end user. This is an important distinction because environmental benefits can be realised within the enterprise, such as through reduced pollution or from material savings, or through use by the end user, in many cases the final consumer. For instance, the environmental benefits of low-energy consumer appliances are realised during their use by the consumer. The introduction to the question also specifies that an environmental innovation can be introduced intentionally, in order to reduce environmental impacts, or can be a side-effect of other innovation goals.

The second question asks about different drivers, including current regulations, expected regulations, grants or other financial incentives, expected demand, and voluntary codes of practice. The final question asks if the enterprise has procedures to identify its environmental impacts.

All questions are asked on a simple 'yes or no' basis. The simple format of the questions resulted from two rounds of cognitive testing with the managers of 20 enterprises.

PACE surveys

Another way to obtain relevant results for eco-innovation is the use of national surveys of pollution abatement and control expenditures (PACE). Since 1996, such surveys have been used on an *ad hoc* basis by several OECD countries (OECD, 2003). In most countries, surveys of this type are limited to firms with more than 20 employees.

Pollution and abatement control activities are defined as "purposeful activities aimed directly at the prevention, reduction and elimination of pollution or nuisances arising as a residual of production processes or the consumption of goods and services" (OECD, 2003, p. 9). This definition excludes unintentional environmental benefits. There are two types of expenditures on these activities: purchase of end-of-pipe technologies and investments in cleaner production technologies (integrated process changes).

A major limitation of PACE data is that they do not differentiate between capital expenditures to purchase innovative technology and expenditures on non-innovative technology to expand production (line extensions). In the latter case, the firm already uses the technology but purchases additional equipment. The PACE survey for the United States covers supporting activities such as innovation expenditures, but these specifically exclude capital expenditures and wages for research.[6]

Box 4.1. Eco-innovation module of the EU's Community Innovation Survey 2008

Innovations with environmental benefits

An environmental innovation is a new or significantly improved product (good or service), process, organisational method or marketing method that creates environmental benefits compared to alternatives.

- The environmental benefits can be the primary objective of the innovation or the result of other innovation objectives.
- The environmental benefits of an innovation can occur during the production of a good or service, or during the after sales use of a good or service by the end user.

During the three years 2006 to 2008, did your enterprise introduce a product (good or service), process, organisational or marketing innovation with any of the following environmental benefits?	**Yes**	**No**
Environmental benefits from the production of goods or services within your enterprise		
Reduced material use per unit of output	□	□
Reduced energy use per unit of output	□	□
Reduced CO_2 'footprint' (total CO_2 production) by your enterprise	□	□
Replaced materials with less polluting or hazardous substitutes	□	□
Reduced soil, water, noise, or air pollution	□	□
Recycled waste, water, or materials	□	□
Environmental benefits from the after sales use of a good or service by the end user		
Reduced energy use	□	□
Reduced air, water, soil or noise pollution	□	□
Improved recycling of product after use	□	□

During 2006 to 2008, did your enterprise introduce an environmental innovation in response to:	**Yes**	**No**
Existing environmental regulations or taxes on pollution	□	□
Environmental regulations or taxes that you expected to be introduced in the future	□	□
Availability of government grants, subsidies or other financial incentives for environmental innovation	□	□
Current or expected market demand from your customers for environmental innovations	□	□
Voluntary codes or agreements for environmental good practice within your sector	□	□

…/…

Box 4.1. Eco-innovation module of the EU's Community Innovation Survey 2008 *(continued)*

Does your enterprise have procedures in place to regularly identify and reduce your enterprise's environmental impacts? (For example preparing environmental audits, setting environmental performance goals, ISO 14001 certification, etc).

☐ Yes: implemented before January 2006

☐ Yes: Implemented or significantly improved after January 2006

☐ No

Source: Eurostat, final harmonised CIS-2008 questionnaire.

Several changes to the PACE surveys would substantially improve their usefulness for measuring eco-innovation. First, the survey questionnaires need to differentiate between capital expenditures for innovative equipment (not previously used by the firm) and expenditures for line extensions. Second, the surveys should collect data on the firm's innovative activities, such as R&D expenditures to reduce and control pollution. Third, the surveys should be harmonised across OECD countries and implemented on a regular basis. This is not currently the case.

Overall evaluation and suggestions for improvement

Surveys on eco-innovation may take the form either of an official, large-scale format or of smaller one-off surveys which focus on a limited region or a set of sectors. Large-scale national innovation surveys in some countries already include a few questions on eco-innovation and have provided information on the prevalence of innovation with environmental benefits. Smaller surveys can investigate aspects of eco-innovation in far greater depth – for example, motivation for and drivers of eco-innovation, its impacts on costs, employment or skills – but their low response rates may reduce confidence in the results.

The eco-innovation module for the CIS 2008 does not cover many issues of importance for measuring eco-innovation as space constraints limited the eco-innovation module to one page. PACE surveys do not differentiate between investment in innovation and line extensions. With some adjustments and harmonisation among OECD countries, PACE surveys could provide a useful vehicle for collecting data on the adoption of eco-innovation and possibly on investment in innovative activities associated with capital expenditures on end-of-pipe and cleaner production technologies. However, it might be difficult to collect information on R&D and other eco-innovation activities through PACE surveys, since many firm managers responsible for capital investments (the target respondent) may not be responsible for innovation.

Organising new surveys dedicated to eco-innovation could help collect in-depth data on different aspects of eco-innovation, particularly unquantifiable information such as the nature of eco-innovation, its drivers and barriers and micro-level effects. In an ideal world, an eco-innovation survey should include questions that are relevant for developing policies that encourage firms to invest in eco-innovation and for informing policy makers of benefits and possible problems, such as the effect of eco-innovation on competitiveness. The following points might be considered for inclusion in future surveys:

- Cover both creative innovation (the enterprise itself invests in developing eco-innovations) and technology adoption (the enterprise purchases relevant technology from external sources) and distinguish between the two.
- Where possible, questions should be asked about R&D investment in eco-innovation, the number of personnel active in research on eco-innovation, and intermediate outputs such as relevant patents.
- Cover different types of eco-innovation (products, processes, marketing, organisational and institutional innovation) to identify where in the value chain and how eco-innovation is occurring.
- Include both intended and unintended eco-innovation to determine where policy incentives should be focused and where they are unnecessary.
- Ask about the types of policies and organisational methods the enterprise has for identifying and correcting environmental impacts. This information is valuable for assessing whether or not these policies make a difference and if so, the sectors on which governments need to focus efforts to encourage more firms to adopt pro-environmental activities.
- Obtain data on the economic effects of eco-innovation on sales, production costs, and employment in order to identify the effects of eco-innovation on competitiveness and possible wider implications for the macro-economy.
- Ask about the appropriation methods used by the firm to benefit financially from eco-innovation.[7]
- Ask about the drivers of eco-innovation, including policies (subsidies, mandates, regulations) and other incentives (exploiting new markets, image, etc.).

It is also useful to obtain some information in relation to a specific innovation such as whether the innovation was introduced in response to a specific policy. As noted above, general questions on drivers or effects are useful, but the design of good policies frequently requires information on the effect of a specific policy on a specific type of innovation and the economic effects of that innovation. Such issues can be addressed by asking respondents to select their most important eco-innovation in terms of its environmental benefits, and by including a series of related questions.[8] It would also be useful to obtain basic data on the environmental impact of the enterprise's products and production processes, although this may be sensitive information.

Where possible, an eco-innovation survey should be linked to official data registers in order to obtain quality information on control variables and on financial information, such as the enterprise's profits, employment and sales over time. In many countries, this is not possible, particularly for surveys by academic bodies. In such cases, the following types of control variables need to be included in the eco-innovation questionnaire:

- firm-level attributes (sector, employment, sales or other output measure).
- commercial conditions (scope of the firm's markets [where and what it sells], level of competition, and if possible, profitability).

Conclusions

Quantitative measurement can be one of the most important ways to better understand eco-innovation, although fully capturing the complex and diverse nature of eco-innovation activities is a challenge. This chapter reviews existing methods of measuring eco-innovation at the macro level in order to understand the strengths and weaknesses of current methodologies and to provide recommendations for improving the metrics available on eco-innovation.

Eco-innovation activities can be investigated from a number of perspectives: the nature of eco-innovation, its drivers and barriers, and its impacts. These aspects can be measured and analysed by using four data categories: input measures; intermediate output measures; direct output measures; and indirect impact measures. Relevant data can be obtained either by using generic data sources or by conducting specially designed surveys.

Table 4.2. Summary of methods for measuring eco-innovation

Mode of measurement	Data sources	Strengths	Weaknesses
Generic data sources			
Input measures	R&D expenditures, R&D personnel, other innovation expenditures (*e.g.* design expenditures, software and marketing costs)	Relatively easy to capture related data	Tend to capture only formal R&D activities and technological innovations
Intermediate output measures	Number of patents, numbers and types of scientific publications	Explicitly provide an indication of inventive output Can be disaggregated by technology groups Combine coverage and details of various technologies	Measure inventions rather than innovations Biased towards end-of-pipe technologies Difficult to capture organisational and process innovations No commonly agreed and applied category for environmental innovations Commercial values of patents vary substantially.
Direct output measures	Number of innovations, descriptions of individual innovations, sales of new products from innovations	Measure actual innovations Timeliness of data Relative ease to compile data Can provide information about types of innovations, *i.e.* incremental or radical	Need to identify adequate information sources Process and organisational innovations are difficult to count The relative value of innovations is hard to identify
Indirect impact measures	Changes in resource efficiency and productivity	Can provide the link between product value and environmental impact Can be compiled at multiple levels: product, company, sector, region and nation Can depict various dimensions of environmental impact	Difficult to cover environmental impact over the entire value chain No simple causal relation between eco-innovations and eco-efficiency

…/…

Table 4.2. Summary of methods for measuring eco-innovation *(continued)*

Mode of measurement	Data sources	Strengths	Weaknesses
Specialised surveys			
Large-scale surveys	EU Community Innovation Surveys, official questionnaire surveys performed regularly, PACE surveys	High response rates Can trace trends in innovation activities over time	Generally can include only a few questions of relevance to eco-innovation PACE surveys are not harmonised among countries; they do not differentiate capital expenditures for eco-innovation from those for line extension.
Small-scale surveys	One-off questionnaire surveys, interviews	Can focus on eco-innovation in far greater depth Possibility to ask about many aspects of eco-innovation	Low response rates Only a few international surveys exist
Panel surveys	Gather information from the same firms over time	Can provide information about size, levels, direction and sources of innovation activities Can identify trends and changes in innovative behaviour over time	Costly to conduct

Each measurement approach has its strengths and weaknesses: no single method or indicator can capture eco-innovation comprehensively. The scope of generic data sources is limited as none is specially designed to measure eco-innovation. For instance, there is no statistical category for eco-innovation in patent databases, R&D statistics or trade journals. Furthermore, generic data sources rarely provide information on the drivers, barriers or impacts of eco-innovation, and most do not provide direct measures of eco-innovation. That said, generic data sources can still yield a wealth of information if more effort is devoted to direct measurement of innovation outputs using documentary and digital sources. Eco-innovation can also be measured indirectly through changes in resource efficiency and productivity. Both of these avenues could usefully receive more attention.To obtain a deeper and broader understanding of eco-innovation, beyond the creation and implementation of end-of-pipe technologies, designing a new dedicated survey or a supplement to an existing survey may prove useful. Surveys can enable researchers to obtain more detailed and focused information on various aspects of eco-innovation such as the type of innovation, drivers and barriers, and micro-impacts. This is especially the case if the survey is conducted internationally on the basis of the same methodology. It would also be useful to conduct panel surveys or interviews to gather information from the same firms over time. Such in-depth surveys may help understand how the nature of eco-innovation is changing and how eco-innovation relates to overall corporate management and performance.

Table 4.2 summaries the strengths and weaknesses of different methods of obtaining data on eco-innovation reviewed in this chapter. In sum, no single method can capture eco-innovation comprehensively. To identify overall patterns of eco-innovation, it is important to apply different analytical methods, possibly combined, and examine information from various sources with an appropriate understanding of the context of the data considered.

Notes

1. This chapter draws primarily on the results from the EC-funded Measuring Eco-Innovation (MEI) project (*www.merit.unu.edu/mei*). René Kemp and Anthony Arundel of Maastricht Economic and Social Research and Training Centre on Innovation and Technology (UNU-MERIT) contributed to this chapter and were involved in the MEI project as project leader and researcher.

2. IMPRESS (Impact of Clean Production on Employment in Europe: An Analysis using Surveys and Case Studies) was led by the Centre for European Economic Research (ZEW), Germany.

3. A trade journal or trade magazine is a periodical, magazine or publication which targets a specific industry or type of trade/business.

4. Part of this section is drawn from Arundel *et al.* (2007).

5. Three very small surveys are excluded from Table 4.1 (Williams *et al.*, 1991; Garrod and Chadwick, 1996; Pimenova and van der Vorst, 2004). Doyle (1992) only surveys environmental equipment manufacturers and is of less interest here.

6. For example, the 2005 PACE survey for the United States (implemented in 2006) states that the survey covers "all related support activities, including but not limited to monitoring and testing and environmentally-related administrative activities", but elsewhere the survey specifically excludes research (DOC, 2005).

7. Appropriation methods refer to strategies companies may employ to protect an innovation against imitation by competitors. Secrecy and intellectual property right protection (patents, licensing) are the most important strategies.

8. This method is widely used by both academic surveys and national survey organisations. For instance, Statistics Canada regularly asks respondents to its innovation survey to identify their most important innovation and to answer a few questions on it. This approach was followed by the IMPRESS study for eco-innovation (Rennings and Zwick, 2003).

References

Andrews, S.K.T., J. Stearne and J.D. Orbell (2002), "Awareness and Adoption of Cleaner Production in Small to Medium-Sized Businesses in the Geelong Region, Victoria, Australia", *Journal of Cleaner Production*, Vol. 10, pp. 373-380.

Arimura, T.H., A. Hibiki and N. Johnstone (2007), "An Empirical Analysis of Environmental R&D: What Encourages Facilities to be Environmentally Innovative?", in N. Johnstone (ed.), *Environmental Policy and Corporate Behaviour*, Edward Elgar, Cheltenham, pp. 142-173.

Arundel, A. and A. Rose (1999), "The Diffusion of Environmental Biotechnologies in Canada: Adoption Strategies and Cost Offsets", *Technovation*, Vol. 19, pp. 551-560.

Arundel, A, R. Kemp and S. Parto (2007), "Indicators for Environmental Innovation: What and How to Measure", in D. Marinova, D. Annandale and J. Phillimore (eds.), *International Handbook on Environment and Technology Management*, Edward Elgar, Cheltenham, pp. 324-339.

Ashford, N. (1993), "Understanding Technological Responses of Industrial Firms to Environmental Problems: Implications for Government Policy", in K. Fischer and J. Schot (eds.), *Environmental Strategies for Industry: International Perspectives on Research Needs and Policy Implications*, Island Press, Washington, DC, pp. 277-307.

Blum-Kusterer, M. and S. Hussain (2001), "Innovation and Corporate Sustainability: An Investigation into the Process of Change in the Pharmaceuticals Industry", *Business Strategy and the Environment*, Vol. 10, pp. 300-316.

Department of Commerce, United States (DOC) (2005) *Pollution Abatement Costs and Expenditures (PACE) Survey Guidelines*, Washington, DC.

Dodgson, M. and S. Hinze (2000), "Measuring Innovation: Indicators Used to Measure the Innovation Process: Defects and Possible Remedies", *Research Evaluation*, Vol. 8, No. 2, pp. 101-114.

Doyle, D.J. (1992), *Building a Stronger Environmental Technology Exploitation Capability in Canada*, report for Environment Canada and Industry, Science and Technology Canada, DSS Contract KE144-1-2273/01-SS, July, Doyletech Corporation, Kanata.

European Commission (EC) (2004), *Stimulating Technologies for Sustainable Development: An Environmental Technologies Action Plan for the European Union*, COM(2004)38Final, Brussels.

EC (2008), *European Innovation Scoreboard 2007: Comparative Analysis of Innovation Performance*, PRO INNO Europe Paper No. 6, Office for Official Publications of the European Communities, Luxembourg.

Frondel, M., J. Horbach and K. Rennings (2004), "End-of-Pipe or Cleaner Production?: An Empirical Comparison of Environmental Innovation Decisions across OECD Countries", ZEW Discussion Paper No. 4-82, ZEW, Mannheim.

Frondel, M., J. Horbach and K. Rennings (2007), "End-of-Pipe or Cleaner Production?: An Empirical Comparison of Environmental Innovation Decisions across OECD Countries", *Business Strategy and the Environment*, Vol. 16, No. 8, pp. 571-584.

Garrod, B. and P. Chadwick (1996), "Environmental Management and Business Strategy: Towards a New Strategic Paradigm", *Futures*, Vol. 28, No. 1, pp. 37-50.

Getzer, M. (2002), "The Quantitative and Qualitative Impacts of Clean Technologies on Employment", *Journal of Cleaner Production*, Vol. 10, pp. 305-319.

Green, K., A. McMeekin and A. Irwin (1994), "Technological Trajectories and R&D for Environmental Innovation in UK Firms", *Futures*, Vol. 26, No. 10, pp. 1047-1059.

Griliches Z. (1990), "Patent Statistics as Economic Indicators: A Survey", *Journal of Economic Literature*, Vol. 28, No. 4, pp. 1661-1707.

Horbach, J. (2008), "Determinants of Environmental Innovation: New Evidence from German Panel Data Sources", *Research Policy*, Vol. 37, pp. 163-173.

Johnstone, N. (ed.) (2007), *Environmental Policy and Corporate Behaviour*, Edward Elgar, Cheltenham.

Kleinknecht A., K. van Montfort and E. Brouwer (2002), "The Non-Trivial Choice between Innovation Indicators", *Economics of Innovation and New Technologies*, Vol. 11, No. 2, pp. 109-121.

Lanjouw, J.O. and A. Mody (1996), "Innovation and the International Diffusion of Environmentally Responsive Technology", *Research Policy*, Vol. 25, pp. 549-571.

Lefebvre, E., L.A. Lefebvre and T. Stéphane (2003), "Determinants and Impacts of Environmental Performance in SMEs", *R&D Management*, Vol. 33, No. 3, pp. 263-283.

Maastricht Economic Research Institute on Innovation and Technology (MERIT) *et al.* (2008), *MEI Project about Measuring Eco-Innovation: Final Report*, under the EU's 6th Framework Programme, MERIT, Maastricht.

Mill, S. and D. Gee (1999), *Making Sustainability Accountable: Eco-efficiency, Resource Productivity and Innovation*, EEA Topic Report, No. 11/1999, European Environment Agency, Copenhagen.

OECD (2003), *Pollution Abatement and Control Expenditure in OECD Countries*, reportfor the Working Group on Environmental Information and Outlooks, Environment Policy Committee, OECD, Paris, *www.oecd.org/dataoecd/41/57/4704311.pdf.*

OECD (2004), *Patents and Innovation: Trends and Policy Challenges*, OECD, Paris.

OECD and Statistical Office of the European Communities (Eurostat) (2005), *Oslo Manual: Proposed Guidelines for Collecting and Interpreting Innovation Data* (3rd ed.), OECD, Paris.

OECD (2008a), *OECD Science, Technology and Industry Scoreboard 2007: Innovation and Performance in the Global Economy*, OECD, Paris.

OECD (2008b), "Preliminary Indicators of Eco-innovation in Selected Environmental Areas", internal working document for the Working Party on National Environmental Policies, Environment Policy Committee.

Pfeiffer, F. and K. Rennings (2001), "Employment Impacts of Cleaner Production: Evidence from a German Study Using Case Studies and Surveys", *Business Strategy and the Environment*, Vol. 10, pp. 161-175.

Pimenova, P. and R. van der Vorst (2004), "The Role of Support Programmes and Policies in Improving SMEs' Environmental Performance in Developed and Transition Economies", *Journal of Cleaner Production*, Vol. 12, No. 6, pp. 549-559.

Rennings, K. and T. Zwick (eds.) (2003), *Employment Impacts of Cleaner Production*, ZEW Economic Studies, Vol. 21, Heidelberg.

Scott, J.T. (2003), *Environmental Research and Development: US Industrial Research, the Clean Air Act and Environmental Damage*, Edward Elgar, Cheltenham.

Steger, U. (1993), "The Greening of the Board Room: How German Companies are Dealing with Environmental Issues", in K. Fischer and J. Schot (eds.), *Environmental Strategies for Industry: International Perspectives on Research Needs and Policy Implications*, Island Press, Washington, DC, pp. 147-166.

Wilén, H. (2008), "Government Budget Appropriations or Outlays on R&D – GBAORD: GBAORD per Inhabitant in the US is Double that of the EU", *Statistics in Focus*, No. 29/2008, Eurostat, Luxemburg.

Williams, H.E., J. Medhurst and K. Drew (1991), " Corporate Strategies for a Sustainable Future", in K. Fischer and J. Schot (eds.), *Environmental Strategies for Industry: International Perspectives on Research Needs and Policy Implications*, Island Press, Washington, DC, pp. 117-146.

World Business Council for Sustainable Development (WBCSD) (2000), *Eco-efficiency: Creating More Value with Less Impact*, WBCSD, Geneva, *www.wbcsd.org/web/publications/ eco_efficiency_creating_more_value.pdf*.

Zutschi, A. and A. Sohal (2004), "Adoption and Maintenance of Environmental Management Systems: Critical Success Factors", *Management of Environmental Quality*, Vol. 15, No. 4, pp. 399-419.

Chapter 5

Promoting Eco-innovation: Government Strategies and Policy Initiatives in Ten OECD Countries

Closer integration of innovation and environmental policies would help achieve ambitious environmental and socio-economic goals simultaneously and benefit from new market opportunities in the growing eco-industry. This chapter reviews existing national strategies and overarching initiatives related to eco-innovation in ten OECD countries (Canada, Denmark, France, Germany, Greece, Japan, Sweden, Turkey, the United Kingdom and the United States) based on responses to a questionnaire survey. The strategies and initiatives are diverse in focus and character and include both supply-side and demand-side measures. A more comprehensive understanding of the interaction between supply and demand will be necessary to create successful policy mixes for promoting eco-innovation in the future.

Introduction

Environmental concerns have gained prominence in the policy arena in the last few decades. The reduction of greenhouse gas (GHG) emissions, for example, has been a top government priority, and many countries have adopted legally binding long-term policy frameworks in order to cut emissions. These frameworks have led to the establishment of a great variety of policy programmes, notably in energy, transport, building and manufacturing.

Like general innovation, eco-innovation needs government interventions that set the right framework conditions and provide enough support for successful research and business development. This chapter takes stock of existing government policy strategies and initiatives intended to promote eco-innovation. Mainly based on responses provided by governments to a specially designed questionnaire survey, existing policy initiatives are considered from the viewpoint of how innovation policy measures are currently utilised to promote eco-innovation.

The chapter starts by briefly outlining the rationale for the integration of innovation and environmental policies and the general status of policy integration. It then reviews existing national strategies and overarching initiatives related to eco-innovation and examines how the concept is defined and which actors have been actively involved in the implementation of such strategies. Next, the existing policy initiatives of the ten governments surveyed are categorised according to a list of innovation policy measures. The chapter concludes with an overview of current policy practices for promoting eco-innovation.

Synergising innovation and environmental policies for eco-innovation

Traditionally, governments in OECD countries have addressed policies for promoting sustainable manufacturing and eco-innovation mainly through their environmental policies. Over the past years, increasing attention is, however, paid to eco-innovation as part of "third-generation innovation policies" by some OECD member countries (OECD, 2005, p. 57).

While sustainable manufacturing and eco-innovation should be embedded in both innovation and environmental policies, these two policy areas have long been separate in OECD countries. The separation is most visible in the fact that these policies have been the responsibility of different ministries. Innovation policy in most countries has been under the ministries for trade and industry and science and technology. Environmental policy has usually been developed by environment ministries. Few efforts have been made to integrate these two policy domains.

Eco-innovation and environmental policy

Before the 1990s, environmental policies tended to be "reactive, informal and often voluntary, based on negotiation between industry and government", with a focus on treatment of industrial wastes. In the 1990s, the concept of integrated pollution prevention and control (IPPC) took hold. Although this approach recognises the importance of technologies for environmental protection, the focus was still largely on end-of-pipe solutions, rather than on the whole production and disposal process (Parliamentary Office of Science and Technology, 2004).

The positive effects of environmental policy on innovation have thus been relatively limited in the past, since stringent regulations and standards do not necessarily provide firms with enough incentive to innovate beyond end-of-pipe solutions. They have however helped to reduce environmental impacts significantly. Moreover, they typically impose greater costs on firms than other policies to reduce environmental impacts (OECD, 2008a). Recently, market-oriented instruments such as green taxes and tradable permits have appeared as more cost-effective measures that put a price on the "bad". However, if eco-innovation is to realise its potential, policies ranging from appropriate investments in research to support for commercialising breakthrough technologies will be needed to ensure that the full cycle of innovation is efficient.

Added to the broad concern regarding traditional instruments of environmental policy is the lack of integration that has been apparent in environmental policy. For example, air and water quality and waste disposal were traditionally tackled independently, making it difficult to identify options for more encompassing initiatives (Heaton, 2002).

Eco-innovation and innovation policy

While environmental policy has been insufficiently oriented towards technology development and innovation, innovation policy has often been too broad to address specific environmental concerns appropriately. Innovation policy has traditionally focused on spurring economic growth by developing new technologies for improving productivity and developing new areas of functionality. This has mainly involved the provision of support to science and technology activities and infrastructure.

Integrating innovation and environmental policies

Eco-innovation has thus not been a main objective of either environmental or innovation policy. Yet a 2005 OECD report on the governance of innovation systems listed a number of benefits to be gained from integrating innovation and environmental policies. From the environmental viewpoint, one benefit would be greater environmental and cost effectiveness. A more innovation-oriented environment policy could more readily improve environmental quality through the application of new technologies, which could also reduce the costs imposed by environmental measures. Second, closer integration could help decouple environmental pressures from economic growth and achieve ambitious environmental and socio-economic goals simultaneously, while benefiting from new market opportunities in the growing eco-industry. From the innovation point of view, it is increasingly recognised that "third generation innovation policies have to become fully horizontal and support a broad range of social goals if they are to achieve their objective of increasing the overall innovation rate in societies" (OECD, 2005; see Box 5.1).

Government strategies for eco-innovation

In order to take stock of information on existing policy initiatives for promoting eco-innovation and to see how each country frames eco-innovation and co-ordinates relevant policies among ministries for policy integration, the OECD conducted a survey of eco-innovation policies through its Committee on Industry, Innovation and Entrepreneurship (CIIE). Responses were received from Canada, Denmark, France, Germany, Greece, Japan, Sweden, Turkey, the United Kingdom and the United States. A summary of their responses is listed in Annex 5.A. Based on these responses, this section presents a brief overview of current government strategies for promoting eco-innovation and describes the government actors involved in the planning and implementation of such strategies.

Countries' views on eco-innovation

There is no consensus on the definition of eco-innovation among the countries surveyed. In some, the term eco-innovation is not used. Commonly used terms include "sustainable manufacturing", "environmental innovation" and "clean-tech" (OECD, 2008b).

Some countries seem to view eco-innovation in a rather traditional sense – the development of environmentally friendly technologies. Canada considers eco-innovation as science and technology work on clean energy research, development, demonstration and deployment. It also refers to the creative process of applying knowledge and the outcome of that process. The US Department of Commerce (DOC) defines "sustainable manufacturing" as the creation of manufactured products that use processes that are non-polluting, conserve energy and natural resources, and are economically sound and safe for employees, communities and consumers.

Box 5.1. Mutually reinforcing links between innovation and environmental policies

There are several good reasons why a more explicitly innovation-oriented environmental policy is needed:

- *Environmental effectiveness:* An innovation-oriented environmental policy is necessary to promote the development and introduction of a new series of techniques that make major improvements in environmental quality more attainable.
- *Decoupling economic growth from environmental pressure*: An innovation-oriented environmental policy is necessary to achieve simultaneously ambitious socio-economic and environmental objectives and substantially raise the eco-efficiency of the economy.
- *Cost-effectiveness*: An innovation-oriented environmental policy is necessary to reduce the cost of environmental measures and achieve more environmental results for the same level of costs.
- *Take advantage of win-win opportunities*: An innovation-oriented environmental policy is necessary to focus on win-win opportunities that have remained unused in order to lower production costs and at the same time pollute less.
- *Market and socio-economic benefits*: An innovation-oriented environmental policy is necessary to benefit from the promising market and socio-economic benefits of the fast-growing environmental industry.

At least three main reasons for a more explicitly environmentally oriented innovation policy can be mentioned:

- *Innovation policy promotes R&D on promising future technologies.* Given the scale and magnitude of environmental problems, technologies limiting the environmental damage of production and consumption are important. Such innovations are not only hampered by "positive" knowledge spillovers that discourage inventors in general but also by "environmental externalities" in the diffusion stage. In such a situation, there is obviously an important role for innovation policy in remediating these market failures.
- *Environmental innovations have some particular properties* compared to most other types of technologies. This is why there is relatively little environmental R&D. First is the importance of government policy in creating demand by regulatory and other environmental instruments. Second is the fact that R&D in environmental innovations is often very complex because it usually involves various scientific and technical disciplines and the necessary competence may not be available in the company undertaking the research.
- *Innovation policy needs to be internalised by other policy domains* to be comprehensive and perform through better integration with the demand side. Innovation becomes a pull factor if it is part of sectoral policies and if public tenders take it explicitly into account. These "third-generation" innovation policies have to become fully horizontal and support a broad range of social goals if they are to achieve their objective of increasing the overall innovation rate in societies.

Source: Dries *et al.* (2005), "Linking Innovation Policy and Sustainable Development in Flanders", in OECD (2005), *Governance of Innovation Systems, Volume 1: Synthesis Report*, OECD, Paris.

Other countries, notably in Europe, have a more encompassing view. Germany's understanding of eco-innovation is not limited to environmental goods and technologies, but includes all technologies, products and services that lead to environmental and economic benefits.[1] It also includes new business models and services (*e.g.* leasing or energy contracting) or consulting activities that lead to environmental and economic benefits. For Greece, eco-innovation extends across all sectors to embrace both technological and non-technological innovations that lead to better environmental performance. While the Environmental Technology Action Plan (ETAP) of the European Union (EU) primarily focuses on accelerating the development of environmental technologies and eco-industries, its definition of eco-innovation also refers to non-technological elements of innovation such as services and management and business methods.[2]

Figure 5.1. The scope of Japan's eco-innovation concept

Target / Field	Industry		Social infrastructure		Personal lifestyle
	Manufacturing	Service	Energy	Transportation / urban	
Technology	· Sustainable manufacturing · Innovative R&D (energy saving, etc.) · Green ICT · Rare metal recycling	· Innovative R&D (Building Energy Management System)	· Innovative R&D (renewable energy, batteries) · Superconducting transmission	· Innovative R&D (intelligent transport systems) · Green automobiles · Maglev	· Heat pump
Business model	· Green procurement (including BtoB) · Green servicizing · EMA · LCA	· Energy services · Environmental rating/green finance	· Green certification	· Modal shift	· Green procurement · Cool biz · Green finance
Societal system (institution)	· Environmental labeling system · Starmark · Green investment		· Top Runner Programme · PRS Act (Renewables Portfolio Standard)	· Green tax for automobiles · Next-generation vehicle and fuel initiative (METI)	· Telework, telecommuting · Work-life balance

Source: Ministry of Economy, Trade and Industry (METI), Japan.

Japan's view is even broader. The government's Industrial Science Technology Policy Committee considers eco-innovation as a concept which provides the direction and vision for societal changes and states that eco-innovation is "a new field of techno-social innovations [that] focuses less on products' functions and more on [the] environment and people" (METI, 2007). In this vision, eco-innovation aims at the development of a sustainable economic society that focuses on reforming not only technologies but also social organisation to ensure minimal environmental impact (Figure 5.1). Japan seeks to develop sustainable production systems and infrastructures that promote zero emissions in order to utilise natural resources and energy in the most efficient way.

Strategies and policy co-ordination

Public awareness of climate change and other environmental concerns is increasing, and most OECD countries emphasise the environment or sustainable development as a top priority in their national strategies. In some of the countries surveyed, eco-innovation is not specifically mentioned, but it seems to be part of their innovation policy and/or environmental policy. Germany, for example, has a clear plan to bridge innovation and environmental policies in its national strategy. This policy integration was the main focus of its 2008 Master Plan for Environmental Technologies, which covers topics such as climate protection, preservation of resources (materials efficiency) and water technologies. In this plan, the integration of environmental policy, innovation policy and other important policy areas is viewed as the way to promote eco-innovation and to open up leading markets for environmental technologies.

The United States clearly recognises the need for "policy innovation" to achieve eco-innovation in industry. In 2002, the Environmental Protection Agency (EPA) established the National Center for Environmental Innovation, which focuses on creating a "results-oriented" regulatory system, promoting environmental stewardship across society and building capacity for innovative problem solving.[3] For the promotion of environmental technologies, research and development (R&D) still attracts a lot of attention and public funding, but there is a clear orientation towards problem solving and focus on commercialisation and dissemination of technologies.

Some countries actively aim to view environmental issues not as a barrier to economic development but as the next opportunity area for innovation, one which would lead to economic growth and greater competitiveness. Although no governmental strategy exclusively addresses eco-innovation, Japan considers that eco-innovation should lead its innovation strategy and the concept has been referred to in several innovation documents such as the recently revised New Economic Growth Strategy (investment in resource efficiency) and the "Innovation 25" Guidelines (strategies for reform of the social system). Greece's Strategic Plan for the Development of Research, Technology and Innovation 2007-13 promotes eco-innovation as a driver for moving the country's economy towards the knowledge economy in 11 thematic priority areas. In 2007 the UK Commission on Environmental Markets and Economic Performance brought together leaders from business, trade unions, universities and non-governmental organisations (NGOs) to develop recommendations on how to exploit economic opportunities arising from the transition to a low-carbon, resource-efficient economy. In July 2009 the government published the UK Low Carbon Industrial Strategy, which sets out the vision for the transition to a low-carbon economy.[4]

France is taking an interesting bottom-up approach to developing national strategies for determining the future course of eco-innovation. *Le Grenelle de l'Environnement* (the Environment Roundtable) was organised in 2007-08 as a nationwide consultation, with the participation of representatives from five stakeholder groups: state, business, trade unions, local authorities and NGOs.[5] Over 30 thematic committees were set up and the participants defined guidelines and objectives for concrete programmes for sustainable development in the fields of housing, transport, renewable energy, waste and recycling, governance, etc. Two bills have been submitted to the National Assembly to ensure the implementation of the outcomes of the roundtable. The first provides main targets and general guidance on implementation; the second defines some compulsory measures as part of the national commitment to the environment (Box 5.2 describes the Dutch government's bottom-up approach in the energy field).

As eco-innovation relates to a number of policy areas, it has been placed under the responsibility of different government departments. Generally, a few government departments are mainly in charge of eco-innovation – typically the ministry of the environment and the ministries for economy and trade and science and technology – with the minor engagement of sector-specific ministries and agencies for energy, natural resources, transport, construction, etc. Such multi-ministerial participation in policies relating to eco-innovation or sustainable development in general is increasing. In the United Kingdom, for example, five departments promote eco-innovation: the Department for Environment, Food and Rural Affairs (Defra), the Department for Transport (DfT), the Department for Communities and Local Government (CLG), the Department of Energy and Climate Change (DECC), and the Department for Business, Innovation and Skills (BIS). In Canada, Industry Canada, Environment Canada, Natural Resources Canada as well as other government departments are involved.

The United States aims for even wider engagement. The DOC's Manufacturing and Services Unit created an interagency working group on sustainable manufacturing under the Interagency Working Group on Manufacturing Competitiveness, which brings together some 17 agencies.[6] For its part, France merged in 2007 the departments responsible for relevant areas into one body for better co-ordination – now called the Ministry of Ecology, Energy, Sustainable Development and Sea.[7]

Box 5.2. The Dutch transition management approach

In order to address the uncertainty and complexity of environmental problems and the interdependence of related policies, the government of the Netherlands adopted a "transition management" approach in its fourth Environmental Policy Plan. This approach sets a long-term vision, which constitutes a framework for formulating future policy objectives and transitional pathways. Interim targets and short-term policies are derived by back-casting from the long-term objectives. This approach also intends to allow policy makers to think in terms of "system innovation" by taking different policy domains into account and engaging different actors.

On this basis, six ministries have been working together to apply this approach to innovation in energy policy, with a view to attaining a sustainable energy supply within 50 years. This Energy Transition Programme first identified seven priority themes (bio-based raw materials, sustainable mobility, chain efficiency, new gas, sustainable electricity, energy in the built environment, and "greenhouse as energy source") for the transition to a sustainable energy system, based on a multi-stakeholder consultation process and scenario studies. For each theme, representatives from industry, academia, NGOs and the government worked together and proposed several paths and experiments. The Energy Transition Task Force, consisting of leading stakeholders, has been working to identify favourable opportunities and specify what needs to be done by the government and others to exploit them. Some of the selected transition experiments are currently under way.

The transition management approach is expected to enable the government to organise its policy around a cluster of options, without choosing specific solutions, while giving an overall policy direction to the market. It also provides opportunities for the government to facilitate networks and coalitions among actors in the transition paths as well as to build mutual trust with stakeholders by sharing common goals.

Source: Reid and Miedzinski (2008); Kemp and Loorbach (2005); Loorbach *et al.* (2008); SenterNovem's Energy Transition website *www.senternovem.nl/energytransition*.

The variety of arrangements raises concerns over horizontality and appropriate co-ordination. The means of implementing policy co-ordination and integration also appear diverse. They range from a centralised approach under a single ministry to a somewhat diffuse networking approach involving many agencies. However, it is not necessarily clear from questionnaire responses whether any ministry plays a clearly leading or co-ordinating role for cross-ministerial collaboration, or whether different ministries are working together effectively on integrating innovation and environmental policies.

Government policy initiatives for eco-innovation

Following the above overview of government strategies, this section reviews public policies and programmes for promoting sustainable manufacturing and eco-innovation in the ten OECD governments that responded to the questionnaire on eco-innovation policies. Existing innovation policies are reviewed, with suggestions on how they could be used to promote a more integrated approach to improving environmental sustainability. This overview of innovation policies provides a basic framework for the following evaluation of current government policy initiatives. The information provided by the survey is supplemented by the "ETAP roadmaps" prepared by EU member states under the ETAP,[8] and by profiles of eco-innovation policies in non-EU OECD countries complied by the OECD Environmental Policy Committee (EPOC).[9]

There are many ways to categorise policy measures relevant to innovation and there so far appears to be no standard taxonomy. For environmental policy the categories have been more clearly established (*e.g.* CSCP *et al.*, 2006). The European Commission (EC)'s *European Innovation Scoreboard*, which monitors innovation policy in EU member states, classifies it into 25 different types of measures (EC, 2008). This classification does not necessarily meet the needs of policy analysis as it is constructed from a statistical perspective. Furthermore, there is growing recognition that many of the problems for promoting innovation arise not only from insufficient investment in innovation activities or inappropriate technologies but also from the lack of relevant markets for innovative products and services. That is, to address innovation more effectively it is necessary to take into account "demand-side" policy measures as well as the traditional "supply-side" measures. None of the measures listed in the *European Innovation Scoreboard* is explicitly oriented towards demand.[10]

Table 5.1. Taxonomy of innovation policy measures

Supply-side measures	Demand-side measures
• Equity support • Research and development • Pre-commercialisation • Education and training • Networks and partnerships • Information services • Provision of infrastructure	• Regulations and standards • Public procurement and demand support • Technology transfer

Source: Adapted from Edler and Georghiou (2007), "Public Procurement and Innovation: Resurrecting the Demand Side", *Research Policy*, Vol. 36.

The taxonomy of supply- and demand-side innovation policy measures of Edler and Georghiou (2007) is primarily applied in this chapter (Table 5.1). Inevitably, there is some overlap between different measures and many policy initiatives also combine several measures as policy mixes. In the following section, the policy initiatives are classified according to their main focus.

Supply-side measures

Equity support

Entrepreneurial activity often involves large commercial and financial risks that cannot always be addressed by market mechanisms alone. Access to finance is often cited as the main constraint on innovation by firms, and public policy has long aimed at easing firms' access to finance. Venture capital funds are one of the major ways of sharing risk through means such as loans, equity injection or participation in management. Another common form of equity support to business is guarantee funds, which guarantee loans to companies directly or indirectly.

This is also the case for eco-innovation. To enable the creation and development of eco-innovative products and green entrepreneurial firms, public policy can implement a variety of equity support for eco-innovation activities and actors. Examples of such financial instruments include: specialised venture capital funds that provide seed capital, green funds to guarantee bank loans for investment projects, and investment guarantee funds that target intermediary financing activities between loans and equity (van Giessel and van der Veen, 2004).

Many governments have taken measures to ease access to finance for firms that develop innovative technologies through venture capital. The focus is often on small and medium-sized enterprises (SMEs), as they suffer most acutely from market failure and find it difficult to obtain funding. However, governments have only introduced a small number of specific measures or instruments for firms developing environmental technologies or eco-friendly products and services, as most equity support measures target general business start-up and development. Some examples with a partial focus on financing for eco-innovation are:

- Denmark: The Danish Investment Fund (*Vaekstfonden*), a government-sponsored investment fund, provides seed and start-up financing to small innovative firms on commercial terms using equity or state-guaranteed loans.[11] Including this fund, 12% of all investments made by venture funds in Denmark went to clean-tech companies in 2007; more than half benefited foreign companies.
- Greece: The Environmental Plans Action provides grants (up to 40% of investment cost) to enterprises for improving their environmental performance as a pre-requisite for certification with an eco-label for their products or the EU Eco-Management and Audit Scheme (EMAS). Support is available for "soft" actions such as testing expenses, certification and consulting services, as well as for process modifications and improvements directly related to environmental areas. During 2000-06, 130 enterprises were selected for funding for a total budget of EUR 16.1 million.

Research and development

R&D policy has long been regarded as the main pillar of innovation and science and technology policies. R&D support programmes are designed to boost innovation activities by directing resources towards a wide range of institutions – universities, basic research institutes, industrial research centres, corporate laboratories and governmental organisations. Government support for R&D is provided either directly, through public research projects, or through the funding of research activities of other public and private institutions.

R&D activities are at the heart of eco-innovation, because they are essential for developing environmental technologies. Although it may be difficult to separate "environmental R&D" from general R&D, public-sector R&D expenditures for "control and care of the environment" represent 5% of total R&D expenditures at most (OECD, 2008c).

In the countries surveyed, most R&D programmes seem to be mainly sector- or technology-specific. In the United Kingdom, new investment is mainly directed towards sustainable energy technology in areas such as off-shore wind and marine energy technology. Canada's Automotive Innovation Fund is an R&D initiative for the automotive sector aimed at developing fuel-efficient vehicles, while energy-related programmes include the Program of Energy Research and Development and the ecoENERGY Technology Initiative (ecoETI). Sweden funds several research programmes and competence centres for different technologies and also has a focus on green nanotechnologies and biotechnologies. The Swedish Energy Agency funds research programmes and competence centres in the fields of renewable energy and energy efficiency. In the United State, the EPA leads the Technology Innovation Programme and the Hydrogen, Fuel Cells and Infrastructure Technologies Programme.

It seems that strategic approaches for shifting the course of entire R&D programmes towards a more environmental or eco-innovation focus are rare. In France, however, Article 19 of the bill on the implementation of *Le Grenelle de l'Environnement* mentions that supportive measures for the transfer and development of new technologies should take account of their environmental performance. Greece indicates that such a shift is mainly driven by a more general restructuring effort towards a competitive economy. In all cases, the proportion of total R&D expenditures directed towards eco-innovation is not clear. Furthermore, R&D for general-purpose technologies, such as information technologies, biotechnologies and nanotechnologies, could be very relevant to eco-innovation, but may not be identified as such. The following examples provide relatively encompassing approaches to "environmental R&D":[12]

- France: The Research Programme on Eco-technologies and Sustainable Development (PRECODD) promotes the development of environmental technologies, including pollution control, as well as new approaches to increasing eco-efficiency in modes of production and consumption.[13] The Environment and Energy Management Agency (ADEME) focuses on supporting SMEs at the early design phase of eco-innovation prior to obtaining R&D funding in three ways: feasibility studies of projects from the technical and economic points of view; use of consultancy services; and temporary appointment of highly qualified personnel for the realisation of the design phase. Article 19 also states that research expenditures for clean technologies and prevention of environmental damage will gradually increase to reach the level of research expenditures for civil nuclear energy by the end of 2012.

- Germany: The "Renewable Resources" funding programme funds R&D and demonstration in the areas of sustainable production of raw materials and energy, environmentally friendly products, and sustainable use of natural resources (forestry and agriculture). The Research for Sustainability Framework Programme promotes the study, implementation and dissemination of innovations for sustainable development with funds of EUR 800 million. Fields of action include: sustainability in industry and business, sustainable use concepts for regions, sustainable use of natural resources and strategies for social action.
- Greece: The country's Strategy for Research Technology and Innovation aims to increase research and technological development expenditure from 0.61% of gross domestic product (GDP) in 2004 to 1.5% in 2015. Eco-innovation appears in most of the thematic areas. Environmental R&D funding will be targeted to actions relevant to climate change, environmental intelligence, risk forecasting and assessment for all types of natural hazards, management of ecosystems and natural resources, and environmental technologies for agricultural pollution, air pollution, water and soil pollution, and solid waste.
- Japan: Environmental R&D efforts focus mainly on energy efficiency, "Green IT", green chemistry, nanotechnologies and new materials through several programmes. The Cool Earth – Innovative Energy Technology Program identified 21 key technologies and created the Map of Technical Strategy.[14] It focuses particularly on the potential contribution of information and communication technologies (ICTs) to higher efficiency in energy and resource use. As a consequence, there is investment in R&D for energy-saving home network technologies, photonic network technologies, high-performance network sub-systems using nanotechnologies, and remote sensing technologies for consistent CO_2 measurement.

Pre-commercialisation

Innovations do not come to the market straight from the R&D stage; there are many stages of innovation from conception of an idea to successful commercialisation as marketable products and services. The EPA, for example, classifies this "R&D continuum" into six stages: research or proof of concept; development; demonstration; verification; commercialisation; and diffusion and utilisation.[15]

Public intervention must take the right form and focus at different stages of this continuum, notably during the demonstration and verification phases, which come just before commercialisation and are increasingly seen as critical. Demonstration involves tests of first-time or early-stage technologies and may be pilot or full-scale. It may involve considerable redesign and de-bugging in order to establish final robustness and optimisation. Verification includes testing of ready-to-market technologies and reporting on their performance to guarantee their quality to users.

Many available environmental technologies have not been successfully introduced into the market, either because the market is not yet well developed or because existing infrastructures and production and consumption systems are an obstacle to their commercialisation (Tukker *et al.*, 2008). Consideration of post-R&D stages in innovation policy is therefore particularly relevant to eco-innovation. In the field of verification, environmental technology verification (ETV) schemes have recently been introduced in Canada, Japan, the United States, etc. to accelerate the entry of new environmental technologies into the marketplace. There has also been international discussion of mutual recognition of different ETV schemes for promoting technology transfer beyond national borders.

Governments have started to recognise the importance of these post-R&D stages of the innovation process. Many initiatives have been introduced to help firms bring newly developed environmental technologies to the market. The current focus of these measures is, however, sometimes limited to promising energy- and transport-related technologies. Examples include:

- Canada: CanmetENERGY is an energy, science and technology organisation working on clean energy research, development, demonstration and deployment with a focus on clean technologies to reduce pollution and GHG emissions.[16] Support for energy technology demonstrations is also provided by Sustainable Development Technology Canada, an organisation that finances and supports the development and demonstration of clean technologies that provide solutions relating to climate change, clean air, water quality and soil. Three Canadian Environmental Technology Advancement Centres also support the development, demonstration and deployment of innovative environmental technologies. They assist SMEs by providing support services, such as general business development counselling, market analysis, assistance in raising capital and technical assistance.
- Denmark: In 2008 the government launched the Energy Technology Development and Demonstration Programme to support the development and demonstration of new efficient energy technologies, including

biomass, wind, solar, fuel cells and hydrogen, as well as technologies for efficient energy use in building, transport and industry.

- France: The Demonstrators Fund was created in July 2008 to support the demonstration of promising environmental technologies in transport, energy and housing, which require testing under real-life conditions. It will provide EUR 400 million between 2008 and 2012 to manufacturers or industry associations which plan demonstration projects with public or private partners ("demonstrators"). The Agency for Innovation and Growth of SMEs (OSÉO) was established in 2005 to provide innovation support and funding to SMEs for technology transfer and innovative technology-based projects with real marketing prospects.[17]

- Japan: The METI runs the New Regional Development Program, which supports a model of "Pioneering Social Systems" to achieve a safe, low-carbon society in its regions by utilising the country's advanced environmental and technological capabilities. In particular, the programme focuses strongly on issues to be dealt with immediately under the two pillars: low carbon emissions and restrained use of natural resources; and safe living. The Eco-innovation Project, under this programme, supports experiments for creating new social systems in various regions in an effort to explore technical "seeds" in local areas by utilising low-carbon technologies.

- Sweden: With the latest budget bill and the Research and Innovation Bill 2009-10, the government shifts its innovation policy focus from grants to technology development and to measures that create markets for energy-efficient, climate-friendly technologies. These bills will allow investing in demonstration plants for second-generation biofuels and other energy technologies, notably those relating to vehicles and electricity production on the verge of commercialisation.

- United Kingdom: The government has been running a number of technology demonstration programmes relating to hydrogen and fuel cell technology, as well as carbon abatement technology. The UK Environmental Transformation Fund focuses specifically on the demonstration and deployment phases of bringing low-carbon and energy-efficient technologies to the market. The Centre of Excellence for Low Carbon and Fuel Cell Technologies focuses on catalysing market transformation projects, linking technology providers and end users.

- United States: The Department of Energy's (DOE) Technology Commercialization Fund (TCF) complements angel investment or early-stage corporate product development (USD 14.3 million in fiscal years 2007 and 2008).[18] The TCF brings the DOE's national laboratories and industry together to identify technologies that are promising but face the "commercialisation valley of death". It makes matching funds available to any private-sector partner that wishes to pursue deployment of the technology identified.

Education and training

As it is people who create new knowledge, many countries have been using education and training programmes to develop skills and talent in order to boost innovation. In the area of education and human capital, innovation policy has tended to focus on the development of science and technology skills, particularly in tertiary education, as graduates with science and engineering degrees have been considered the most valuable inputs into the R&D process. Public policies for education have also started paying attention to linking higher education and business and introducing programmes to nurture entrepreneurship among students.

As in other innovation-related policy areas, education and training programmes are critical to eco-innovation. They develop the human capital needed to deliver eco-innovative solutions and create a potential labour force for "green jobs". The provision of tailored programmes on eco-innovation thinking or environmental issues in general could help to create future environmental researchers and engineers. It could also drive innovation in a more sustainable direction, if students embraced the environment as an integral part of future societal development. Education and training would also be relevant to demand-side policy, as it generally builds public concern for environmental challenges and helps shift consumer behaviour to a more sustainable mode.

A review of governmental policies shows that governments are aware of the need to develop skills and unlock talent to unleash the innovation potential necessary to meet strategic societal challenges. The United Kingdom, for example, has set itself to be a world leader in skills in the context of the Leitch Review on long-term skills needs.[19] Support for education and training in the countries surveyed has also involved awareness-raising programmes. Several countries have taken measures to mainstream environmental education in the school curricula or vocational training:

- Denmark: Technology service institutes were established as independent knowledge bodies to deliver knowledge to enterprises. They plan to include climate change issues in vocational training.
- Germany: Programmes for raising teaching skills to cope with the environmental and sustainability challenges are being introduced in vocational training for agricultural occupations.
- Greece: The country's regional Centres of Environmental Education offer targeted environmental education programmes for students, employees and teachers.
- Sweden: The Law for Higher Education introduced in 2006 states that universities have a responsibility to promote sustainable development in the curricula.

While most OECD countries aim at upgrading skills in general and introducing sustainability issues in their curricula, a few focus on specific skills for eco-innovation. The following initiatives recognise the importance of creating a knowledgeable workforce in emerging environmental industries:

- Canada: ECO Canada was set up with government funding as a not-for-profit education and employment organisation which focuses on environmental training directed by industry and its stakeholders.[20] Its mission is to ensure an adequate supply of people with the skills and knowledge required to meet the environmental human resource needs of the public and private sectors. Human Resources and Skills Development Canada has provided financial support for many of ECO Canada's projects.
- United States: The EPA has organised a wide range of programmes of environmental education and training. The Green Act authorised funding to establish national and state job training programmes to help American workers apply for jobs in the renewable energy and energy-efficiency industries. The Energy Independence and Security Act authorised the creation of the Energy Efficiency and Renewable Energy Worker Training Program to train for "green collar" jobs. The Green Engineering Program developed a textbook entitled *Green Engineering*, which can be used by educators to promote green thinking in engineering processes and applications.[21] This programme also developed continuing education courses for engineers working in industry.

Networks and partnerships

Innovation policy has recently come to better recognise the fact that innovation diffuses through knowledge networks. Knowledge creation is a sophisticated, dynamic process, and many innovations come not only from corporate R&D centres or in-house innovation programmes but also from user groups, consumer networks and supplier channels outside the firm (von Hippel, 2005). The concept of "open innovation" exemplifies that innovation takes place in a world of networks and a web of relations in which firms have to participate (Chesbrough, 2006).

Until recently, innovation programmes were mostly project-based and targeted particular groups of researchers. In recognition of the significance of networks and partnerships for innovation, many policy programmes have sought to influence the structure of innovation by requiring co-operation in research projects and supporting network development. Van Giessel and van der Veen (2004) consider that "the spillovers of government intervention increased and the effects of a subsidy programme became longer lasting than the projects of the programme".

The review of the concepts and examples of sustainable manufacturing and eco-innovation in Chapters 1 and 2 clearly highlight the importance of knowledge networks in the creation of eco-innovative solutions, particularly for closed-loop production and more service-oriented provisions. In order to improve the overall sustainability performance of products and services, innovation activities need to address the entire value chain, notably through life cycle assessment. Here, government has a role to play as a facilitator of networks grouping diverse innovation actors, notably through public-private partnerships and networking platforms for eco-innovation.

Many OECD countries recognise the importance of knowledge networks and have extensively embedded the support of such networks in their innovation policy. For most of the countries surveyed, today's environmental challenges require a new approach to policy making that fosters eco-innovation through collaboration. To date, there are a few successful networks specifically targeted at developing new environmental technologies and solutions (Box 5.3 describes EU initiatives in this area).

Box 5.3. EU platforms for eco-innovation

The European Commission has several initiatives for establishing platforms and networks composed of expert stakeholders to help fulfil the Lisbon Strategy objectives of "a competitive Europe":

- **European technology platforms** bring together stakeholders, led by industry, to define medium- to long-term research and technological development objectives and better align EU research priorities with industry needs. Over 35 sector- or technology-specific platforms have been launched, including several in environmental technology areas such as wind energy, sustainable mineral resources, renewable heating and cooling, sustainable chemistry, and zero-emission fossil fuel power plants. In each platform, participating stakeholders are expected to go through the following three-stage process collectively:
  - agree upon a common vision for the technology;
  - define a strategic research agenda setting out the necessary medium- to long-term objectives for the technology;
  - implement the strategic research agenda by mobilising significant human and financial resources.

Instead of focusing on a specific sector or technology, the Manufuture Technology Platform was established to take a horizontal approach to engaging a broad spectrum of industries. It aims to develop a strategy for research and innovation that makes it possible to speed up the rate of industrial transformation to high-added-value products, processes and services that fit the future knowledge-driven economy, including eco-efficient products and new business models. So far, this platform has developed a "Common Vision towards 2020" and trans-sectoral technology roadmaps, and has set up 30 national and regional initiatives.

- **The Competitiveness and Innovation Programme – Entrepreneurship and Innovation Programme (CIP-EIP)** supports projects in eco-innovation through three initiatives: financial instruments, networks of actors, and pilot and market replication projects. Under networks of actors, Europe INNOVA was launched by DG Enterprise and Industry in 2006 to identify and analyse the drivers and barriers to innovation in specific sectors by bringing together public and private providers of support for innovation. In the first phase of its Sectoral Innovation Watch project (2006-08), eco-innovation was investigated as a horizontal topic along with sectoral themes (see Reid and Miedzinski, 2008). In 2009, a new set of actions was launched to establish two European Innovation Platforms – on clusters and eco-innovation – and to reinforce the European Innovation Platform on Knowledge-Intensive Services. The platforms aim to test innovative tools through public-private partnerships with a view to leveraging their broader deployment in priority sectors, such as those of the Lead Market Initiative and of the ETAP. Eco-innovation is not an exclusive topic of the Platform on Eco-innovation; a project on renewable energies (KIS-PIMS) is already running on the Platform on Knowledge-Intensive Services, and the Platform for Clusters indicates energy efficiency and eco-innovation as suitable sectors in its call for proposals.

Source: European Technology Platform website: *www.cordis.lu/technology-platforms*; Manu*future* Technology Platform website: *www.manufuture.org*; Europe INNOVA website: *www.europe-innova.org*.

- United Kingdom: The Technology Strategy Board (TSB) in charge of promoting technology-driven innovation relies heavily on networking to drive innovation among UK businesses. It has set up:

  - Innovation platforms, to pull together policy, business, government procurement, research perspectives and resources to generate innovative solutions to societal issues and harness the innovative capabilities of UK businesses.[22] Innovation platforms focus on particular areas of innovation in order to identify available levers and funding streams, including two innovation platforms in the environment-related areas of low-impact buildings and low-carbon vehicles. For example, the Low Carbon Vehicle Innovation Platform will provide GBP 40 million to support R&D and commercialisation of low-carbon vehicles.

  - Knowledge transfer networks (KTNs), to increase the depth and breadth of transfer of professional knowledge into UK-based businesses.[23] Networks exist in the fields of technology and business application, including environmental fields such as resource efficiency and fuel cells. KTNs bring together people from business, universities, research, finance and technology organisations to stimulate innovation through knowledge transfer.

  The TSB conducted a major review which confirmed the value of the KTNs: 75% of business respondents rated KTN services as effective, 50% developed new R&D and commercial relationships with people met through these networks, and 25% made a change to their innovation activities as a result of their engagement. The most highly rated functions of the KTNs are monitoring and reporting on technologies, applications and markets, providing quality network opportunities, and identifying and prioritising key innovation-related issues and challenges. In view of the increasingly global nature of innovation, there will be an increase in the support given by KTNs to international activities.

- United States: The Green Suppliers' Network was established by the EPA in collaboration with the DOC to help small and medium-sized manufacturers stay competitive and profitable, while reducing their impact on the environment.[24] It works with large manufacturers to engage their suppliers in low-cost technical reviews to identify strategies for improving process lines and using materials more efficiently. The "lean and clean" initiative aims at eliminating non-value-added activity to drive down costs and improve efficiency in the manufacturing process. It has a particular emphasis on the elimination of industrial wastes.

Furthermore, government initiatives for sustainable development are increasingly implemented in collaboration with industry and local actors, sometimes through formal public-private partnerships, notably in the area of town planning, housing, transport, etc. Such partnership-based innovation programmes include:

- Denmark: The government has created five partnerships to strengthen innovation in Danish enterprises in areas such as water and industrial biotechnologies. Their goal is the development of new business concepts and competitive eco-efficient technological solutions.
- France: "Competitiveness clusters" have been established since 2004 in various regions to conduct innovative projects focused on one or more identified markets in partnership between businesses, research institutes and training organisations.[25] Several of the existing 71 clusters are currently implementing joint environmental technology projects with high growth potential either in renewable energy and energy efficiency or in a specific sector. Examples include decentralised energy (Languedoc-Roussillon),[26] chemistry and the environment (Rhône-Alpes),[27] industry and agro-resources (Champagne-Ardennes),[28] city and sustainable mobility (Île-de-France)[29] and vehicles of the future (Alsace and Franche-Comté).[30] Such initiatives are expected to bring growth and employment opportunities to the regions and increase the attractiveness of France through enhanced international visibility.
- Germany: A series of programmes have been financing national and international co-operative ventures between SMEs and research establishments, or innovation clusters and interlinking activities.
- Greece: Through a combination of EU, public and private funds, five regional "innovation poles" were established between 2000 and 2006 to promote co-operation among industry, enterprises, academia and research centres. Two of the regional innovation poles focus on environmental priorities. "SynEnergia" in West Macedonia promotes innovation in environmental management of power plants, biomass, hydrogen and renewable energy technologies.[31] The West Greece Pole focuses among other things on the management of industrial wastes and natural resources.
- Japan: The Eco Town Programme set up in 1997 requires municipalities or regions to develop an Eco Town development plan for local resource circulation with comprehensive involvement of industry and citizen groups. The plan should reflect the area's specific characteristics and advantages. By 2006, 26 towns had been approved as

Eco Towns and subsidies provided for both hardware and software projects.[32]

- Sweden: The Delegation for Sustainable Cities was formed in 2008 to support initiatives and projects from local authorities and the business sector in the area of sustainable city building.

Information services

Provision of information also plays a basic role in helping businesses build technological competences and obtain customer-oriented knowledge that can support their innovation activities. It can also help them keep up to date on legislation and international standards. Information centres that collect and provide up-to-date information on technological and business developments constitute one of the most common instruments. They can help close information gaps among firms, particularly SMEs, which often suffer from a lack of access and resources for obtaining latest technological know-how. Information centres may be operated by public authorities or by private organisations such as chambers of commerce and professional contractors, possibly with public funding. They may have physical offices around the country or may only exist virtually through websites (CSCP *et al.*, 2006).

The government also plays an essential role in diffusing knowledge and information on environmental issues and eco-innovation. To foster eco-innovation, information centres can be designed to provide information and promote transfer of knowledge on resource efficiency and environmental technologies. These functions can be complemented by knowledge exchange networks and education and training programmes as well as consulting services.

Responses to the questionnaire show that this area has yet to be developed in many countries, as most advisory services for SMEs have not specifically targeted environmental issues, let alone eco-innovation. Information for firms on environmental issues has mainly been provided through the Internet. Existing environment-focused information services include:

- Canada: Information services provided through websites by the government include "Funding Technologies for the Environment", an inventory of funding and incentive programmes to help develop, demonstrate and deploy environmental technologies.[33]
- Denmark: The Danish Technological Institute provides a web portal to give enterprises easy access to the latest knowledge on biotechnology, ecology, environmental chemistry, energy, materials and food.[34]

- France: The ADEME helps SMEs to adopt environmental management methods in both their production and products by: undertaking an eco-audit, or obtaining an ISO 14001 or EMAS certification; and designing or improving products at each stage of their life cycle.[35]
- Germany: An Internet portal, "Cleaner Production Germany", provides comprehensive information about the performance of German environmental technologies and services.[36]
- Japan: The Energy Conservation Center, Japan, a foundation which aims to promote the efficient use of energy and sustainable development and protect against global warming, provides a website for the industrial, civil and transport sectors with information on energy conservation and Top Runner product standards.[37]
- Turkey: The Technology Development Foundation informs SMEs on phasing out the use of ozone-depleting substances in different sectors and technology alternatives.[38]
- United Kingdom: The government funds the Energy Saving Trust which provides free information and advice and has a network of local advisory centres throughout the country specifically designed to help companies and consumers take action to save energy.[39]
- United States: The EPA created the Environmental Technology Opportunities Portal to match companies and organisations with programmes for fostering environmental technologies and to relay information on EPA's technologies for air, water and waste treatment and control.[40] The DOC's Sustainable Manufacturing Initiative and Public-Private Dialogue established a web portal for companies which provides information on what the DOC and other federal agencies are doing to support sustainable manufacturing.[41]

Provision of infrastructure

In recent years, policy makers and researchers have begun to consider certain types of infrastructure as a crucial support for innovation activities. In particular, transport and communications infrastructure has increasingly been viewed as essential for economic success and for raising productivity. Transport factors such as commuting time and proximity to market can play a prominent role in a region's capacity to attract companies and talents. A high-speed digital network now improves a region's ability to innovate, to attract entrepreneurs and to create demand for digitally based products and services. Digital network technology has even allowed many businesses to conduct their operations and produce their products in a new and innovative way. In some industries, access to natural resources can be important for innovation (State of Minnesota, 2008). The increasing policy focus on industry clustering for inducing innovation and creating competitive advantages has also led to the provision of better infrastructure to particular areas as well as the creation of science and industry parks.

The provision of infrastructure is also very important for sustainable manufacturing and eco-innovation. Needless to say, if ICTs are to be utilised to help reduce CO_2 emissions by reducing transaction costs and controlling manufacturing processes, high-speed broadband access is necessary. Innovation related to vehicles using alternative fuels, user-friendly public transport or renewable energy relies on infrastructure for new fuelling systems, sophisticated traffic control, diffused energy distribution systems, etc. The creation of eco-industrial parks (see Chapter 1) can also be an attractive way for governments to encourage businesses to work together to find innovative solutions for improving resource efficiency and to develop environmental technologies.

Despite the importance of infrastructure provision, it is not yet at the centre of the innovation policies of the countries reviewed.[42] Countries that include ICT infrastructure in eco-innovation measures can be seen as pioneers:

- Denmark: The government established the Action Plan for Green IT in 2008 under the Ministry of Science, Technology and Innovation.[43] The aim is to promote greener ICT use among citizens, businesses and public authorities and to stimulate smart ICT solutions to bring about a reduction in overall energy consumption.
- France: A Green IT consultation group was established in January 2009 to make ICTs less polluting and to promote their use for the development of eco-friendly businesses. The group plans to publish a strategy for encouraging emerging environmentally responsible solutions with the help of the ICT sector and to facilitate the uptake of these solutions by companies, especially SMEs. It estimates that better exploitation of Green IT opportunities would result in growth of 0.5% in the national economy.[44]
- Japan: The government considers the establishment of "zero emissions-based infrastructures" in energy supply, transport and town development to be critical for realising eco-innovation and a sustainable society. In 2008, the METI launched the Green IT Initiative to develop innovative ICTs in a medium- and long-term perspective.[45] Focus areas include infrastructures and technologies for teleworking, intelligent transport system (ITS), home energy management system (HEMS) and building energy management system (BEMS).

Demand-side measures

Demand-side policies aim at market development and typically focus on the end of the innovation cycle, close to business. In the case of eco-innovation, the market has not automatically generated enough innovation effort. One reason may be the lack of a sufficiently large market. Demand approaches to encourage eco-innovation include regulation and standards, public procurement and policies to foster technology transfer as well as a range of other measures.

Regulations and standards

Traditionally, industry tended to view environmental regulations negatively – as an additional cost, as distorting incentive structures and hence as having an adverse effect on competitiveness. It is nonetheless increasingly recognised that regulations – under certain conditions and if designed and implemented properly – can help to create a market for new eco-friendly products and services.

What is required to foster eco-innovation is an appropriately planned, yet flexible regulatory framework. Although the balance is difficult to strike, forward-looking regulations and standards, based on the best available techologies or the overall environmental performance of products or companies, could guide the course of innovation and accelerate the creation of eco-innovative solutions by creating a "level playing field". It is also important for policy makers to encourage and ensure industry's adoption and implementation of regulations and standards by setting an appropriate reporting and monitoring framework. Flexible and well-designed regulations and standards can also encourage the diffusion of advanced environmental technologies and eco-friendly products and services by creating demand for these.

A review of government policies in this area shows that there are many environmental regulations and standards but that most are not necessarily designed to drive innovation for sustainable solutions. Yet, some regulations and standards are appearing which aim at stimulating sustainable manufacturing and eco-innovation by creating demand within firms and among consumers. Most governments surveyed also have eco-labelling schemes to stimulate consumer demand for eco-friendly products:

- Canada: The recent Federal Sustainability Act requires the development of strategies that include goals and targets for sustainable development as well as implementation strategies for the federal government. The Energy Efficiency Act sets minimum energy performance standards for energy-using products such as appliances, lighting and heating, and air conditioning products. Amendments to the act will either set a minimum energy performance standard for a series of new products or will make existing standards more stringent for others. The amendments will come into force between 2007 and 2010. EcoLogo is North America's largest and most respected environmental standard and certification mark. It was founded by the Government of Canada in 1998 and is now recognised worldwide. EcoLogo certifies environmental leaders in over 120 product and service categories, thereby helping customers identify sustainable products.

- Japan: The government has set up a number of new-generation regulations and standards. The METI's Top Runner programme is unique in setting performance targets for enterprises.[46] It adopts a dynamic process of setting and revising standards by taking, in principle, the current highest energy efficiency rate of products as a benchmark standard in 21 product groups. This flexible standard-setting creates positive incentives and competition among manufacturers to quickly improve their product performance and does not call for financial support. The programme is supplemented by the e-Mark voluntary labelling scheme to facilitate consumer choice at the point of sale. To improve corporate environmental management, the Ministry of the Environment set the Environmental Reporting Guidelines and provides awards to acknowledge corporate efforts.[47] To spread environmental awareness among SMEs, which have fewer resources and less capacity, it developed Eco Action 21 in 1996, an environmental management system designed for SMEs.[48]
- United States: The Energy Independence and Security Act signed in December 2007 sets standards to increase energy efficiency and the availability of renewable energy. Its three key provisions are: *i)* the Corporate Average Fuel Economy (CAFE) standards, which target 35 miles per gallon for the combined fleet of cars and light trucks by 2020; *ii)* the Renewable Fuel Standard (RFS), which sets renewable fuel use at 9 billion gallons in 2008 and at 36 billion gallons by 2022; and *iii)* appliance and lighting efficiency standards.

Public procurement and demand support

The public sector is a large consumer: in the EU15, for example, approximately 16% of GDP is spent on public procurement (EC, 2004). Public procurement therefore is a key source of demand for firms, particularly in such sectors as construction, health care and transport. Green or sustainable public procurement has been promoted by many OECD countries since the 1990s as part of environmental policy. However, it has not been mainstreamed in as many countries as expected owing to the higher costs or longer payback periods of many eco-friendly products and services, lack of knowledge among procurement officers, and concerns over potential distortion of fair competition.

For many years, public procurement was not considered a key means of leveraging innovation or part of innovation policy. As attention to demand-side policies gradually increases, some governments have started to highlight procurement as a way to spur innovation (Edler and Georghiou, 2007). The EC issued a strategic innovation policy paper that sheds light on the importance of public procurement for innovation and for creating a lead market (EC, 2006). In 2007 it published a guide for using public procurement to drive innovation (EC, 2007).

Edler and Georghiou (2007) argue for revitalising public procurement as an eco-innovation policy tool with three main rationales: to generate or maintain effective demand for new environmental goods and services; to address structural failures and inefficiencies affecting translation of needs into functioning markets for eco-innovative products; and to raise the quality of public infrastructure and services through up-to-date innovative solutions.

A certain number of the countries surveyed have listed public procurement as a driver of eco-innovation. Little evidence is so far available on the extent of the procurement initiatives and on their success in creating new eco-innovative solutions or lead markets:

- Canada: The Federal Policy on Green Procurement of 2006 uses procurement as a tool to advance innovative environmental technologies and solutions.[49] The policy defines environmentally preferable goods and services as those with a lesser or reduced environmental impact over their life cycles, in comparison with competing goods or services serving the same purpose. Environmental performance considerations include, among other things: the reduction of GHG emissions and air contaminants; improved energy and water efficiency; reduced waste and reuse and recycling; the use of renewable resources; reduced hazardous waste; and reduced toxic and hazardous substances. The policy is expected to increase demand for environmentally preferable goods and services and promote further innovation in the area of environmental technologies.

- Germany: The High-Tech Strategy for Germany attaches importance to boosting the role of state governments in promoting demand for innovation.[50] A web portal has been created to inform decision makers on possibilities of green procurement.

- Japan: The Law on Promoting Green Purchasing of 2000 made it obligatory for all governmental institutions to implement green procurement, and local authorities and private companies were encouraged to adopt green procurement as well.[51] This government initiative was largely influenced by the Green Purchasing Network (GPN) set up in 1996, a multi-stakeholder network with some 3 000 member organisations, including 2 300 companies.[52] The GPN encourages green procurement by all parties in order to create demand for eco-products by establishing product databases and sharing the experience of civil society and industry.

- United States: As one of the world's largest consumers, the US government can potentially provide a strong incentive for eco-innovation. Since 1993, it has aimed to strengthen federal agencies' environmental, energy and transport management. It requires federal agencies to apply sustainable practices when acquiring goods and services, including the acquisition of bio-based, environmentally preferable, energy-efficient, water-efficient and recycled-content products. Both the EPA and the General Services Administration work to help agencies find environmentally preferable products through online guidance[53] and the *Global Supply Environmental Products Catalog.*[54] The Energy Independence and Security Act also promotes the purchase of energy-efficient products and alternative fuels by federal agencies. The Federal Electronics Challenge promotes agencies' purchase of electronics that meet certain environmental criteria.[55]

A "forward commitment procurement" model aims to address the lack of market pull for innovation by providing the market with advance information on future needs in outcome terms. Procurers agree with suppliers to purchase a product or service that currently does not exist, at a specified future date, providing it can be delivered at agreed performance levels and costs. Such a product is expected to effectively solve a specified challenge with an environmental footprint smaller than current solutions.[56] The UK Department for Business, Innovation and Skills (BIS) is supporting public procurers to apply this model. The UK government also plans to establish a "centre of expertise in sustainable procurement" which will help develop new and innovative ways for sustainable working, planning and procurement in the civil service.

Procurement measures can be also applied for business-to-business trade. Government may also directly support business and individual consumers with subsidies, tax incentives or other benefits for purchasing particular eco-products and services such as renewable energy, energy-efficient electronics and green buildings in order to stimulate demand. Notable examples of proactive use of demand support measures to shift the course of technology and product development include:

- France: The *Bonus-Malus* (reward-penalty) scheme was introduced for personal cars in December 2007.[57] This scheme provides a subsidy (EUR 200 to 5 000) or a penalty (EUR 200 to 2 600) to any buyer of a new car depending on the model's amount of CO_2 emissions per kilometre. The emission levels will decrease by five grams of CO_2 every two years, and the introduction of an annual tax instead of the current one-time penalty is being discussed.[58] The extension of the scheme to other household equipment is also under consideration.

  Other green fiscal measures based on the proposal of *Le Grenelle de l'Environnement* include: zero-interest loans of up to EUR 30 000 for financing thermal renovations of houses; tax credits for the interest on loans for acquiring accommodations in line with the prevailing standard; "eco-charges" for heavy trucks; exemptions from the property tax for farms using solar-powered electricity.[59]

Technology transfer

Technology transfer is the process of transferring technologies, know-how, knowledge or skills from one party to another. It often refers to the export of technological competences from industrialised to developing countries, but it can also refer to domestic or local transfer, for example from large companies to SMEs. Public policy can encourage the transfer of promising technologies to local firms through adequate incentive mechanisms or direct intervention. For countries exporting technologies, policy intervention can help expand the market for particular environmental technologies abroad. Policy measures for technology transfer include bilateral or multilateral agreements, working with international development co-operation agencies, establishment of technology transfer institutions, promotion of foreign direct investment, use of export credit, and support for pilot projects.

Successful technology transfer programmes in the area of environmental technologies and know-how is a way for importing countries to increase resource efficiency relatively quickly. At the same time, it can also give exporting countries considerable market and innovation opportunities. However, environmental technologies may not be directly transferable; their adaptation to socio-cultural conditions in recipient countries and the training and engagement of local people are often necessary for successful transfer (CSCP *et al.*, 2006).

Box 5.4. Non-technology transfer for eco-innovation

Under its Cleaner Production Programme, the United Nations Industrial Development Organization (UNIDO) has been working to transfer the service-oriented "chemical leasing" business model to developing countries in order to put closed-loop manufacturing into practice. Completely different from the conventional sales model, in this new business model the customer pays for the benefits obtained from the chemical, not for the substance itself. The supplier does not simply provide the chemical but instead sells the functions and associated know-how on its optimised use, while remaining responsible for the chemical during its whole life cycle: its use, recycling and disposal. Since payment is calculated on the result of functional units (*e.g.* number of pieces cleaned, amount of area coated) instead of the amount of chemicals purchased, the supplier has a strong incentive to reduce the amount of chemical use and the customer will benefit from lower cost.

With support from the government of Austria, where early experiments with this model were made, UNIDO carried out initial knowledge transfer projects in Russia, Mexico and Egypt based on trilateral partnerships between chemical supplier companies, user companies and local National Cleaner Production Centres (NCPCs). For example, an Egyptian chemical supplier, Dr Badawi Chemical Work, started providing GM Egypt, one of its user companies, with services for cleaning with hydro-carbon solvent for a fixed fee per vehicle, instead of selling the solvent per litre. This has led to a reduction of solvent consumption from approximately 1.5 litres per vehicle to 1 litre per vehicle as well as a cost reduction of 15% from savings on raw material use. It has also had environmental benefits in terms of increasing the recycling rate, better management of solvent waste and a more efficient cleaning process.

With positive results from the first pilot projects, this programme has been extended to other countries such as Colombia, Germany, Morocco, Serbia and Sri Lanka.

Source: UNIDO's Chemical Leasing website *www.chemicalleasing.com*; and Sena (2007), "Chemical Leasing and Chemical Management Services", presentation at the International Institute for Industrial Environmental Economics (IIIEE) Network Conference, 28 September, Lund, Sweden.

The countries surveyed take different approaches to technology transfer. While the United States targets India and China as future export markets, Sweden aims to encourage both imports and exports of environmental technologies to expand its market (Box 5.4 gives an example of business model transfer):

- Sweden: The government set up a new export platform, SymbioCity, to market Swedish green technologies and sustainable construction worldwide.[60] It brings together 700 Swedish companies involved in green technologies, sustainable construction and urban planning, and targets foreign cities that want to introduce sustainable development in their planning. Meanwhile, the government has tasked the Invest in Sweden Agency with promoting incoming foreign investments in the clean-tech areas including bio-energy, environmental engineering, green chemistry, heating, ventilation and air-conditioning (HVAC), sustainable building, and waste and recycling.[61]

- United States: The federal government focuses on the creation of markets abroad as a way to export US environmental technologies through technology transfer and international partnerships. The EPA, for example, supports the promotion of exports in clean, efficient energy technologies to India, China and other developing countries. Two initiatives support exports in clean technologies: the Clean Energy Technology Export Program and the Environmental Exports Program. The former is a public-private partnership for addressing export barriers in the world clean technology market; the latter helps mitigate risk for US companies and offers competitive financing terms to international buyers for the purchase of US environmental goods and services.

Conclusions

Traditionally, governments in OECD countries have mainly used their environmental policies to promote sustainable manufacturing and eco-innovation, without necessarily building coherence and/or synergies with other policies. More recently, environmental concerns have started to be integrated in innovation policies. This trend needs to be supported, as environmental and innovation policies can reinforce each other.

From the perspective of policy integration, this chapter reviews national strategies and overarching initiatives related to eco-innovation and examines how the concept is defined and the actors actively involved in implementation of such strategies. It also categorises existing policy initiatives according to a list of innovation policies which includes both supply-side and demand-side measures, and analyses the extent of the integration of innovation and environmental policies.

Results from the questionnaire survey show that an increasing number of countries rightly perceive environmental challenges not as a barrier to economic development but as an opportunity to achieve economic growth and competitiveness through innovation. However, not all countries surveyed seem to have a specific strategy for eco-innovation and when they do, policy co-ordination among different government agencies is limited.

Policy initiatives and programmes introduced by countries to promote eco-innovation are various and include both supply-side and demand-side measures. Measures in support of supply include increased access to finance for firms developing new technologies, funding for R&D and pre-commercialisation, and support for education and training. As most countries surveyed recognise the need for a collaborative approach to developing the technologies needed to face today's environmental challenges, many government programmes in support of supply involve the creation of networks, platforms or partnerships that engage business, academia, government representatives and other stakeholders such as environmental action groups. Most initiatives are organised around a specific sector or technology and non-technological aspects of eco-innovation have often not yet been taken into account.

Demand-side measures, such as green public procurement, regulatory instruments and technology transfer are receiving increasing attention with the recognition that the existence and expansion of relevant markets for innovative products and services is also essential to meet environmental challenges. Yet, it seems that demand-side measures need a more focused approach in order to leverage industry activities for eco-innovation. A more comprehensive understanding of the interaction between supply and demand for eco-innovation – and of the relation between production and consumption of eco-innovative products and services – will be needed to create successful eco-innovation policy mixes.[62] Moreover, better evaluation of the implementation of different sets of eco-innovation measures would be helpful to identify promising eco-innovation policies as well as appropriate contexts in which specific policy instruments can be deployed effectively.

Notes

1. This includes technologies to improve efficiency (energy and material efficiency), sustainable energy generation (especially renewable energy, but also more environmentally friendly energy generation from fossil fuels), waste reduction and treatment technologies, water and wastewater treatment and sustainable water management, technologies and concepts, products and technologies for a sustainable mobility.
2. The ETAP defines eco-innovation as "the production, assimilation or exploitation of a novelty in products, production processes, services or in management and business methods, which aims, throughout its life cycle, to prevent or substantially reduce environmental risk, pollution and other negative impacts of resource use (including energy)".
3. *www.epa.gov/innovation*.
4. *www.berr.gov.uk/files/file52002.pdf*.
5. *www.legrenelle-environnement.fr*.
6. *www.manufacturing.gov/interagency/interagency.asp?dName=interagency*.
7. *www.developpement-durable.gouv.fr*.
8. The national roadmap of each EU member state can be downloaded from h*ttp://ec.europa.eu/environment/etap/policy/roadmaps_en.html*.
9. The issues concerning the integration of innovation perspectives into environmental policy were investigated by the EPOC. The country profiles are available from *www.oecd.org/environment/innovation/globalforum* (OECD, 2008b).
10. The EC has also created the European Inventory of Research and Innovation Policy Measures, which collects and classifies national information and documentation on research and innovation policies, measures and programmes (*www.proinno-europe.eu*). This inventory lists 38 innovation policy instruments in five categories, and includes both supply- and demand-side measures. This categorisation was not used here owing to the large number of categories and for consistency with other OECD work.
11. *www.vaekstfonden.dk*.
12. The EC's seventh framework programme for research and technology development (FP7) for 2007-13 also includes "environment (including climate change)" as one of ten thematic areas for funding of collaborative research.

13. PRECODD was replaced in January 2009 by ECOTECH, a new programme with similar objectives.
14. *www.meti.go.jp/english/newtopics/data/pdf/CoolEarth_E_revised.pdf.*
15. *www.epa.gov/etop/continuum.html.*
16. *http://canmetenergy.nrcan.gc.ca.*
17. *www.oseo.fr.*
18. *www1.eere.energy.gov/commercialization/technology_commercialization_fund.html.*
19. *www.hm-treasury.gov.uk/leitch_review_index.htm.*
20. *www.eco.ca.*
21. *www.epa.gov/oppt/greenengineering/pubs/textbook.html.*
22. *www.innovateuk.org/ourstrategy/innovationplatforms.ashx.*
23. *www.ktnetworks.co.uk.*
24. *www.greensuppliers.gov.*
25. *www.industrie.gouv.fr/poles-competitivite.*
26. *www.pole-derbi.com.*
27. *www.axelera.org.*
28. *www.iar-pole.com.*
29. *www.advancity.eu.*
30. *www.vehiculedufutur.com.*
31. *www.innopolos-wm.eu.*
32. *www.meti.go.jp/policy/recycle/main/english/3r_policy/ecotown.html.*
33. *www.ic.gc.ca/eic/site/fte-fte.nsf/eng/home.*
34. *www.dti.dk.*
35. *http://www2.ademe.fr/servlet/KBaseShow?sort=-1&cid=96&m=3&catid=17579.*
36. *www.cleaner-production.de.*
37. *www.eccj.or.jp.*
38. *www.ttgv.org.tr.*
39. *www.energysavingtrust.org.uk.*
40. *www.epa.gov/etop.*

41. *http://trade.gov/competitiveness/sustainablemanufacturing/index.asp.*
42. Recent stimulus packages to address the economic crisis contain a wider range of measures in this area, however.
43. *www.itst.dk/filer/Publications/Action_plan_for_Green_IT_in_Denmark/index.htm.*
44. *www.secteurpublic.fr/public/article.tpl?id=15360.*
45. *www.meti.go.jp/english/policy/GreenITInitiativeInJapan.pdf.*
46. *www.eccj.or.jp/top_runner/index.html.*
47. *www.env.go.jp/policy/j-hiroba/PRG/pdfs/e_guide.pdf.*
48. *www.env.go.jp/policy/j-hiroba/PRG/pdfs/e_eco_action.pdf.*
49. *www.tpsgc-pwgsc.gc.ca/ecologisation-greening/achats-procurement/politique-policy-eng.html.*
50. *www.bmbf.de/pub/bmbf_hts_lang_eng.pdf.*
51. *www.env.go.jp/en/laws/policy/green/index.html.*
52. *www.gpn.jp.*
53. *www.epa.gov/epp.*
54. *www.gsa.gov/Portal/gsa/ep/contentView.do?P=FXA1&contentId=9845&contentType=GSA_OVERVIEW.*
55. *www.federalelectronicschallenge.net.*
56. *www.dius.gov.uk/policy/public_procurement.html.*
57. *www.developpement-durable.gouv.fr/article.php3?id_article=2825.*
58. One estimate indicates that the sales of cars emitting less than 130g CO2/km during 2008 increased by 46% from the previous year and now represent 45% of the total sales volumes (30% in 2007). On the other hand, the sales of cars emitting over 160g CO2/km dropped to 14% of the total sales volumes in 2008 from 24% in 2007 (Lianes, 2009).
59. *www.developpement-durable.gouv.fr/IMG/pdf/Presentation_des_mesures_fiscales_cle02291f.pdf.*
60. *www.symbiocity.org.*
61. *www.isa.se/templates/Normal____62875.aspx.*
62. It should be noted that the results presented in this chapter are preliminary as the number of governments participating in the questionnaire survey is limited and the information provided does not necessarily cover all relevant policy initiatives in respondent countries.

References

Chesbrough, H. (2006), *Open Business Models: How to Thrive in the New Innovation Landscape*, Harvard Business School Press, Boston, MA.

Dries, I., P. van Humbeek and J. Larosse (2005), "Linking Innovation Policy and Sustainable Development in Flanders", in OECD (2005).

European Commission (EC) (2004), *Buying Green!: A Handbook on Environmental Public Procurement*, Office for Official Publications of the European Communities, Luxembourg.

EC (2006), *Putting Knowledge into Practice: A Broad-Based Innovation Strategy for the EU*, COM(2006)502 final, 13 September, European Commission, Brussels.

EC (2007), *Guide on Dealing with Innovative Solutions in Public Procurement: 10 Elements of Good Practice*, Commission staff working document, SEC(2007)280, Brussels, *www.proinno-europe.eu/doc/procurement manuscript.pdf*.

EC (2008), *European Innovation Scoreboard 2007: Comparative Analysis of Innovation Performance*, PRO INNO Europe paper No. 6, Office of Official Publications of the European Communities, Luxembourg.

Edler, J. and L. Georghiou (2007), "Public Procurement and Innovation: Resurrecting the Demand Side", *Research Policy*, Vol. 36, pp. 949-963.

Giessel, J.F. van and G. van der Veen(eds.) (2004), *Policy Instruments for Sustainable Innovation*, Technopolis, Amsterdam.

Heaton, G.R. (2002), *Policies for Innovation and the Environment: Toward an Arranged Marriage*, paper presented at the six countries programme conference "Innovation Policy and Sustainable Development: Can public innovation incentives make a difference?", 28 February-1 March, Brussels, *www.6cp.net/downloads/02brussels_heaton.doc*.

Hippel, E. von (2005), Democratizing Innovation, MIT Press, Cambridge, MA.

Kemp, R. and D. Loorbach (2005), "Dutch Policies to Manage the Transition to Sustainable Energy", in F. Beckenbach *et al.* (eds.), *Jahrbuch Ökologische Ökonomik 4: Innovationen und Nachhaltigkeit*, MetropolisVerlag, Marburg, pp. 123-150.

Lianes, A. (2009) "Bonus-malus et prime à la casse: ce qu'il faut savoir", *DD Magazine*, 11 February, *www.ddmagazine.com/926-voiture-bonus-malus-ecologique-prime-a-la-casse-conditions-dattribution.html.*

Loorbach, D., R. van der Brugge and M. Taanman (2008), "Governance in the Energy Transition: Practice of Transition Management in the Netherlands", *International Journal of Environmental Technology and Management*, Vol. 9, No. 2/3, pp. 294-315.

Ministry of Economy, Trade and Industry (METI), Japan (2007), *The Key to Innovation Creation and the Promotion of Eco-innovation*, report by the Industrial Science Technology Policy Committee of the Industrial Structure Council, METI, Tokyo.

OECD (2005), *Governance of Innovation Systems, Volume 1: Synthesis Report*, OECD, Paris.

OECD (2008a), "The Economics of Climate Change", internal working document for the preparation of the Meeting of the Council at Ministerial Level.

OECD (2008b), "National Approaches for Promoting Eco-innovation: Country profiles of eight non-EU OECD countries", internal working document for the Working Party on Global and Structural Policies, Environmental Policy Committee.

OECD (2008c), "Encouraging Sustainable Manufacturing and Eco-innovation", internal working document for the Committee on Industry, Innovation and Entrepreneurship.

Parliamentary Office of Science and Technology, United Kingdom (2004), *Environmental Policy and Innovation*, Postnote, No. 212, January, Parliamentary Office of Science and Technology, London, *www.parliament.uk/documents/upload/POSTpn212.pdf.*

Reid, A. and M. Miedzinski (2008), *Sectoral Innovation Watch in Europe: Eco-innovation*, final report to Europe INNOVA initiative, May, Technopolis Belgium, Brussels, *www.technopolis-group.com/resources/downloads/661_report_final.pdf.*

Sena, A.A. (2007), "Chemical Leasing and Chemical Management Services", presentation at the Institute for Industrial Environmental Economics (IIIEE) Network Conference, 28 September, Lund.

State of Minnesota, United States (2008), "Physical Infrastructure: Detailed Explanation of Metrics and Sources", CareerOneStop website, *www.careeronestop.org/Red/AssetMapping/PhysicalInfrastructure.aspx*.

Tukker, A. *et al.* (2008), *System Innovation for Sustainability 1: Perspectives on Radical Changes to Sustainable Consumption and Production*, Greenleaf Publishing, Cheltenham.

UNEP/Wuppertal Institute Collaborating Centre on Sustainable Consumption and Production (CSCP), Wuppertal Institute (WI) and German Agency for Technical Co-operation (GTZ) (2006), *Policy Instruments for Resource Efficiency: Towards Sustainable Consumption and Production*, GTZ, Eschborn.

Annex 5.A

Government Policies and Programmes for Eco-innovation: Country Survey Responses

Canada

Definition and strategy for eco-innovation

Definition of eco-innovation

- Eco-innovation refers to science and technology work on clean energy research, development, demonstration and deployment.
- It also refers to the creative process of applying knowledge and the outcome of that process.
- Innovation can be promoted systematically across the economy, not only in R&D laboratories.

Strategy and initiatives for promoting eco-innovation

- The Federal Sustainable Development Act mandates the development of a national sustainable development strategy. The strategy will be formulated by 2010.
- Several non-profit organisations such as Canadian Environment Technology Advancement Centres and Sustainable Development Technology Canada were created by the government and served to contribute to successful eco-innovation.
- There are several specific strategies and programmes such as: ecoACTION (including ecoTRANSPORT, ecoENERGY and ecoAGRICULTURE programmes), Sustainable Development Technology Canada (SDTC), Industry Canada's Sustainable Development 2006-2009, Industry Canada's Science and Technology Strategy, Canada's Sustainable Cities, Going for the Green, and Technology Roadmaps.

Environment in innovation policies

Overall priorities

- Climate change, clean air, soil, biofuel, and the development of technologies for bio-energy, gasification, carbon capture and storage, electricity transmission, distribution and storage, solar and wind power, and fuel cells.

Supply-side measures

Equity support

- SDTC is a foundation to finance and support the development and demonstration of clean technologies on climate change, clean air, water quality and soil.
- Industry Canada hosts Funding Technologies for the Environment database that lists funding initiatives.
- CanmetENERGY is an organisation that acts as a window to federal financing for developing energy-efficient and clean technology.
- Other relevant governmental funds include: the Automotive Innovation Fund, the Freight Technology Demonstration Fund, and ecoENERGY retrofit funding.

Research and development

- There are several R&D initiatives: Program of Energy Research and Development, Technology and Innovation Research and Development, ecoENERGY Technology Initiative, CanmetENERGY.
- The Canada Foundation for Innovation funds universities and research institutions to carry out world-class research and technology development in the field of renewable resources and environmental research.

Pre-commercialisation

- Canadian Environmental Technology Advancement Centres support the development, demonstration and deployment of innovative environmental technologies. This is done by assisting SMEs through a provision of support services such as: general business development counselling; market analysis; assistance in raising capital; and technical assistance.
- SDTC supports the development and demonstration of clean technologies providing solutions to issues of climate change, clean air, water quality and soil.
- The Environmental Technology Verification System provides a verification of environmental performance claims associated with projects and technologies.

Education and training

- ECO Canada is an organisation that provides environmental training directed by industry and its stakeholders.
- The EcoTechnology for Vehicle programme provides consumer education on low-emission vehicles.

Networking and partnership

- Industry Canada, Environment Canada, Natural Resource Canada and other departments and the private sector collaborate for innovation and environment programmes.
- There are several networks initiatives including: Network of Centres of Excellence, Centres of Excellence for Commercialization and Research, Business-led Networks of Centres of Excellence , Industrial Research and Development Internship Program, Asia-Pacific Partnership Building and Appliance Task Force (partnership of national governments on energy efficiency).

Information services

- Funding Technologies for the Environment is an inventory of funding and incentive programs to help develop, demonstrate and deploy environmental technologies.
- ecoENERGY for Fleets programme provides information and advice for reducing emissions from commercial fleets.

Demand-side measures

Regulation and standards

- The Energy Efficiency Act regulates the energy-use standards of any imported and inter-provincially traded energy-using products, labelling of energy-using products, and collection of data on energy use.

Public procurement and demand support

- The Federal Policy on Green Procurement uses procurement as a tool to advance innovative environmental technologies and solutions.

Co-ordination for eco-innovation

Policy co-ordination within government

- Government departments collaborate and co-ordinate activities for climate change by using regulatory approaches, funding programmes, market-based instruments and awareness raising.
- Industry Canada generally leads the promotion of innovation and facilitates investment in new technologies.

Denmark

Definition and strategy for eco-innovation

Definition of eco-innovation

- Uses the definition of the EU Environmental Technology Action Plan (ETAP): "the production, assimilation or exploitation of a novelty in products, production processes, and services or in management and business methods, which aims, throughout its life cycle, to prevent or substantially reduce environmental risk, pollution and other negative impacts of resource use (including energy)".

Strategy and initiatives for promoting eco-innovation

- The government set the Action Plan for Eco-efficient Technology in 2007 to contribute to solving the global environmental challenge.
- The government launched a Business Climate Strategy in 2009 that aims to combine economic growth with GHG emissions reduction. This strategy is being developed under the guidance of the Business Panel on Climate Change consisting of ministers, business representatives and academics.
- The Action Plan for Green IT was set up by the Ministry of Science, Technology and Innovation in 2008 to promote greener IT use among citizens, business and public authorities, and to promote smart IT solutions that help bring about a reduction in overall energy consumption.

Environment in innovation policies

Overall priorities

- One of main targets is to keep global climate change within a 2°C rise by reducing CO_2 emission by 20-30% by 2020 based on a global agreement.
- The Action Plan for Eco-efficient Technology focuses on nine initiatives: partnerships for innovation; targeted and enhanced export promotion; research and technology development; strengthened efforts to promote eco-efficient technology by the Ministry of the Environment; targeted promotion of eco-efficient technology in the

EU; climate and energy technology; reducing environmental impacts from livestock farms; a clean and unspoilt aquatic environment; a healthy environment.

- The government's energy proposal up to 2025 includes: 100% independent fossil fuels; minimum 30% of renewable energy use; efficient utilisation of energy with an average energy saving of 1.4% between 2010 and 2025.

Supply-side measures

Equity support

- Environment Billon Fund will distribute grants to at least 30 enterprise-based projects for eco-efficient technologies by 2010.
- Clean-tech is one of the focus areas for the state-backed Danish Growth Fund.

Research and development

- A great increase in spending for publicly financed research to 1% of GDP by 2010.
- 2007-09 strategic research projects support R&D in the areas of climate change, energy, water, air pollution, chemical and soil contamination.
- The government will double public funding for research into energy technologies to DKK 1 billion a year by 2010.

Pre-commercialisation

- The Energy Technology Development and Demonstration Programme (EUDP) was launched in 2008 to support development and demonstration of new efficient energy technologies including biomass, wind, solar, full cells and hydrogen as well as technologies for efficient energy use in building, transport and industry.

Education and training

- Universities in Denmark are research-based and some grants for environment-related research can spill over to research-based education.
- Climate change issues are included in vocational training.
- Increase the number of PhD scholarships to 2 400 by 2010.

- An innovation pilot scheme promotes employment of highly qualified staff in SMEs.

Networking and partnership

- Innovation consortia and ICT interaction projects strengthen public sector opportunities to support enterprise innovation.
- Industrial PhD Initiative supports research students who divide their time between working at an enterprise and studying.
- Promoting interaction between academic and research institutions and many enterprises through high-technology networks, regional technology centres and ICT competency centres.

Information services

- The Pesticide Plan 2004-09 provides subsidies to national centres of Danish agriculture to advise farmers on reducing pesticide use.
- The Approved Technological Service (ATS) Institute provides a portal for enterprises to gain easy access to the latest knowledge on biotechnology, fire, ecology, environmental chemistry, energy, materials, food, etc.
- Developing new innovation-promoting instruments for SMEs.

Demand-side measures

Regulation and standards

- Provide consumer advice and promote eco-labelling.
- Danish red Ø logo was made for labelling organic food products.

Technology transfer

- An agreement was signed with China on innovation projects for eco-efficient technologies.

France

Definition and strategy for eco-innovation

Definition of eco-innovation

Eco-innovation does not have a strict definition but could be understood as below:

- In a narrow sense, it means innovation on technologies directly linked to environmental protection (= "innovation in environmental technologies").
- In a more general sense, it means innovation corresponding to the development and/or adoption by one or more organisations of technological or organisational changes in the production of goods and services, or even in the use and treatment at their end of life of products, with a view to better preservation of the environment and improved efficiency in the use and conservation of energy and natural resources with a life cycle approach (= "innovation in eco-responsibility of economic and social actors"). It includes the areas of innovation referred to by terms such as process technologies, product/service, eco-industries, business models, marketing methods, and organisational/institutional changes.

Strategy and initiatives for promoting eco-innovation

- *Le Grenelle de l'Environnement* (Environment Roundtable) was organised in 2007-08 as a nationwide consultation and debate with the participation of five categories of stakeholder representatives: state, business, trade unions, local authorities and NGOs.
- The bill on the implementation of Le Grenelle de l'Environnement sets medium- to long-term national objectives: reduce GHG emissions by 75% between 1990 and 2050 by reducing releases by 3% a year on average; increase the share of renewable energy to at least 23% by 2020; reduce energy consumption of existing buildings by at least 38% by 2020.
- The National Strategy for Sustainable Development will be updated in 2009 under the aegis of the General Commissioner for Sustainable Development.

- The Strategic Committee of Eco-industries was established in 2008. It consists of business leaders and well-known personalities in industry and environmental technologies. A strategic study on the potential of those activities was completed. The future ECOTECH 2012 plan will be based on the committee's proposal.

Environment in innovation policies

Overall priorities

- Le Grenelle de l'Environnement set up 33 thematic committees to define guidelines and objectives for operational programmes in the fields of housing, transport, low-carbon vehicles, research, renewables, waste management and recycling, emerging risks, governance, CSR, etc.
- In its strategic plan 2007-10, the Environment and Energy Management Agency (ADEME) defines ten main areas for financing and developing research and technological innovation activities, including air, buildings, noise, climate change, waste, energy, renewable energy and raw materials, environmental management, sites and soils, and transport.

Supply-side measures

Equity support

- Since 1997, Mutual Funds for Innovating Enterprises (FCPI) have provided private investors with a tax reduction of up to EUR 6 000. From 2008, some funds (FCPI-ISR) focus on financing socially responsible investing enterprises.

Research and development

- The National Research Agency (ANR) and ADEME run the Research Programme on Eco-technologies and Sustainable Development (PRECODD) which promotes the development of environmental technologies, including pollution control, as well as new approaches to increase eco-efficiency in modes of production and consumption. PRECODD was replaced by the ECOTECH programme in January 2009.
- ADEME supports SMEs at the design phase of eco-innovation prior to obtaining funding for development through: feasibility studies of projects from the technical and economic perspective; use of consul-

tancy services; temporary appointment of qualified personnel to carry out the design phase.

- ANR has run programmes dealing with sustainable energy and environment. ADEME also manages, finances and develops research and technological innovation in energy and the environment.

- Article 19 of the bill on the implementation of Le Grenelle de l'Environnement establishes a process and objectives for research for sustainable development. The government will mobilise a supplementary EUR 1 billion by 2012 for research on climate change, energy, future engines, biodiversity, health and the environment, etc. Research expenditures for clean technologies and prevention of environmental damage will gradually increase to reach the level of research expenditures for civil nuclear energy by the end of 2012.

Pre-commercialisation

- The Agency for Innovation and Growth of SMEs (OSÉO) was established in 2005 to provide innovation support and funding to SMEs for technology transfer and innovative technology-based projects with real marketing prospects.

- The Demonstrators Fund was created in July 2008 to support the demonstration of promising environmental technologies in transport, energy and housing, which require experiments under real-life conditions. It will provide EUR 400 million between 2008 and 2012 to "demonstrators".

Education and training

- ADEME helps SMEs to adopt environmental management methods from both production and product perspectives through: an eco-audit or ISO 14001/EMAS certification; designing or improving products at each stage of their life cycle.

Networking and partnership

- The Green IT consultation group was established in January 2009 to make use of ICTs less polluting and to encourage the development of eco-friendly businesses through ICTs.

Demand-side measures

Regulation and standards

- Eco-labelling of consumer goods for informed choices.

Public procurement and demand support

- *Bonus-Malus* (reward-penalty) scheme was introduced for personal cars in December 2007. It provides a subsidy (EUR 200-5 000) or a penalty (EUR 200-2 600) to any buyer of a new car depending on CO_2 emissions per kilometre.
- Diverse green fiscal measures proposed in Le Grenelle de l'Environnement have been implemented: zero interest up to EUR 30 000 for financing thermal renovations of houses; tax credit for the interest on loans for acquiring accommodations in line with the "BBC" standards; "eco-charge" for heavy trucks; exemption of property tax for farms using solar-powered electricity.

Technology transfer

- Article 19 mentions that support measures for the transfer and development of new technologies should take into account their environmental performance.

Co-ordination for eco-innovation

Policy co-ordination within government

- In 2007, the departments responsible for the environment, energy, housing, transport, and land planning were merged into one ministry. It is now called the Ministry of Ecology, Energy, Sustainable Development and Sea in charge of green technologies and climate change negotiations.
- The Ministry of Economy, Industry and Employment co-ordinates the stimulation of eco-industries with ADEME, ANR and OSÉO.
- Promoting eco-innovation requires the integration of policies in favour of sustainable development and the integration of environmental concerns into different policy instruments and innovation projects.

Germany

Definition and strategy for eco-innovation

Definition of eco-innovation

- Eco-innovation is not confined to environmental goods and efficient technologies, sustainable energy generation, waste reduction and treatment technology, but also includes business models, services and consulting activities which bring environmental and economic progress.

Strategy and initiatives for promoting eco-innovation

- The National Strategy for Sustainability was formulated in 2002. A progress report was issued in 2008. The German Chancellery is responsible for this strategy.
- The government created a specific strategy, Masterplan for Environmental Technologies, which was approved in November 2008. The main action fields include water technologies, technologies for materials efficiency and climate protection technologies. The main focus is promotion of the application of eco-innovations and opening up leading markets for environmental technologies.
- The High-Tech Strategy for Germany is the central innovation strategy. In the present legislative period, the federal government is focusing in particular on stimulating research and technology in areas of key importance, including cross-cutting technologies such as biotechnology and nanotechnology as well as energy and environmental technologies. The aim is to build bridges between research and future markets.
- The Integrated Energy and Climate Programme (IEKP) was adopted in 2007 to improve energy efficiency, expand the use of renewable energy and reduce GHG emissions.
- The national ETAP process: The national ETAP network organises the exchange of experiences and develops recommendations for action in line with the German ETAP roadmap. There is a focus on SMEs to improve their access to research, financing tools and global markets.

Environment in innovation policies

Overall priorities

- A study, Roadmap Environmental Technologies 2020, was conducted to develop political and strategic options for future research funding of environmental technologies.

Supply-side measures

Research and development

- The research, technological development and demonstration funding programme aims to develop environmental technologies such as biomaterials and bioenergy from renewable resources.
- A programme on promoting innovation in the fields of nutrition, agricultural and consumer protection was set up in 2006. Grants are made towards R&D projects that are to achieve environmental improvement in agriculture, forestry and fisheries.

Education and training

- The rules on initial and further vocational training in agricultural occupations address rising ecological and sustainability challenges and ensure a sustainable approach to commercial activity.
- The government supports teaching subjects relating to environmental protection and sustainability in further education and lifelong learning.

Information services

- Energy advice programmes provide special funds for raising energy efficiency in SMEs and energy advice for residential properties on the spot or through consumer advice centres.

Demand-side measures

Regulation and standards

- The renewable energies act, co-generation act, renewable energies heat act, and act for opening up metrology for electricity and gas to competition were promulgated in 2007.
- An energy-saving ordinance and rules on the expansion of the electricity grid were revised in 2008.

- A motor tax for passenger vehicles according to level of pollutant and CO_2 emissions.

Public procurement and demand support

- Guidelines on the procurement of energy-efficient products and services were published.

Greece

Definition and strategy for eco-innovation

Definition of eco-innovation

- Eco-innovation is any form of innovation aiming at significant and demonstrable progress towards the goal of sustainable development by reducing impacts on the environment or achieving a more efficient and responsible use of natural resources, including energy.
- Eco-innovation also includes any form of environmentally friendly innovative actions in all sectors that contribute to a substantial improvement in competitiveness, development, employment and citizens' welfare.

Strategy and initiatives for promoting eco-innovation

- The Strategic Plan for the Development of Research, Technology and Innovation for 2007-2013 promotes innovation as a key driver for the transition to the knowledge economy and improvement of competitiveness.
- The Greek National Strategy for Sustainable Development, approved in 2002, aims at economic development while safeguarding social cohesion and environmental quality in the areas of climate change, air pollutants, solid waste, water resources, desertification, biodiversity and natural ecosystems, and forests.
- The operational programme "Competitiveness 2000-2006" aims to promote eco-innovation and environmental investments by enterprises.
- Support for individual businesses in all sectors to receive a ISO 14001 certification for environmental management systems.

Environment in innovation policies

Overall priorities

- The Strategic Plan was formulated around two main priorities: an increase in and improvement of investments in knowledge and excellence towards sustainable development; promotion of innovation, dissemination of new technologies, and entrepreneurship to generate economic and social benefits.

- This plan focuses on 11 priority thematic areas: ICT; farming; food and biotechnology; eco-friendly products and processes in traditional sectors such as textile and construction; advanced materials; nanosciences and microelectronics; energy; transport; environment and health; space and safety engineering; cultural heritage; and social and economic dimension of development. In most of these priority areas, environmental performance improvements are the main area for actions that will be financed.

Supply-side measures

Equity support

- The Environmental Plans Action provides grants to enterprises that implement environmental plans leading to eco-labelling or EMAS certification.
- The Management and Reuse of Industrial Wastes Action provides grants for the creation or expansion of waste management and utilisation plants.
- The Investment Incentives Law, which is the main regional state aid instrument, provides the highest level of grants to enterpises for introducing and adapting to environmentally friendly technologies in the production process or for adopting best available techniques according to the EU IPPC Directive.
- Several other actions within Competitiveness 2000-2006 provided funds to SMEs for investment in equipment replacement, information technologies and certification of management systems, etc.
- The Ministry for Development in co-operation with the Ministry of Economy is planning specific granting schemes for enterprises to make improvements in their environmental performance.

Pre-commercialisation

- The Centre for Renewable Energy Sources (CRES) is the national agency for promoting renewable energy and energy savings. It provides services for measurements of renewable energy technologies' operating characteristics (such as wind turbines and photovoltaics), operates testing laboratories for renewable energy technologies, and is involved in demonstration projects.

Research and development

- Research funding includes the environment as one priority area with an objective to develop environmental intelligence, to manage risk by establishing comprehensive monitoring and prevention approaches, to support indigenous development of the environmental industry, etc.

Education and training

- The National Plan for the Implementation of the UNECE Strategy for Education for Sustainable Development (ESD) was drafted.
- The Regional Centres of Environmental Education offer targeted environmental education programmes for students, employees and teachers.

Networking and partnership

- Through a combination of EU and public and private funds, several science and technology parks and business incubators for knowledge-intensive enterprises have been developed.
- Five regional innovation poles were established in 2000-06 to promote co-operation between industry, enterprises, academia and research centres. Two of the poles focus on environmental protection priorities: SynEnergia in West Macedonia promotes innovation in environmental management in power production plants, biomass, hydrogen and renewable energy technologies; the West Greece Pole focuses, among other things, on management of industrial wastes and natural resources.

Information services

- CRES was established as a national agency for promoting renewables and energy savings.

Demand-side measures

Regulation and standards

- More than 1 000 companies participate in the Collective Alternative Management Scheme for recycling of packaging, used tires, end-of-life vehicles, electric and electronic equipment, batteries, accumulators, lubricant waste and construction waste.

- Other measures include: GHG emission permits for enterprises, implementation of the IPPC directive, eco-labelling, and EMAS certification.

Co-ordination for eco-innovation

Policy co-ordination within government

- The National Research and Technology Council, the Intergovernmental Committee and the National RTD Management Organization were established to co-ordinate government activities related to research policy.

Japan

Definition and strategy for eco-innovation

Definition of eco-innovation

- Eco-innovation is for founding a sustainable economic society by reforming technical innovation and creating a social system that ensures minimum impact on the environment.
- The Industrial Science Technology Policy Committee defines it as "a new field of techno-social innovations that focuses less on products' functions and more on the environment and people".

Strategy and initiatives for promoting eco-innovation

- The Cool Earth 50 Initiative launched by the former Prime Minister targets a reduction of GHG emissions by half by 2050 from the current level.
- The New Economic Growth Strategy revised in 2008 has three pillars: construction of new economic and industrial structures in the era of "resource productivity competition"; reconstruction of a strategy to capture global markets for sustainable development; future-oriented vitalisation of regions, SMEs, agriculture and services.

Environment in innovation policies

Overall priorities

- The Cool Earth – Innovative Energy Technology Program identified 21 key energy technologies and created the Map of Technical Strategy.

Supply-side measures

Research and development

- R&D projects focus particularly on new applications of ICTs, including the development of energy-saving home network technologies, photonics network technologies, high-performance network sub-systems using nanotechnologies, and remote sensing technologies for CO_2 consistency measurement.

Pre-commercialisation

- The Regional Demonstration Project for Global Innovation Architectures provides grants for or commissions demonstrations for exploring technical "seeds" that promote eco-innovation and tackle climate change in local areas.
- METI's New Regional Development Program aims to realise a safe and low-carbon emission society in regions through a model of "Pioneering Social Systems" and to capitalise on the country's strengths in environmental technology capabilities.
- The Environmental Technology Verification (ETV) programme was established to verify the performance of advanced technologies by third parties in the areas of air pollution and water.

Networking and partnership

- METI and the Ministry of the Environment (MoE) implement the Eco Town Program since 1997. It encourages local municipalities, businesses and citizens to work together towards a sound material-cycle society.

Information services

- The Energy Conservation Center, Japan (ECCJ), a foundation which aims to promote the efficient use of energy, protect against global warming and foster sustainable development, provides a website for the industrial, civil and transport sectors to gain access to information on energy conservation and Top Runner product standards.

Provision of infrastructure

- METI launched the Green IT Initiative in 2008 to develop innovative IT technologies with a medium- and long-term perspective. Focus areas include teleworking, intelligent transport system (ITS), home energy management system (HEMS) and building energy management system (BEMS).

Demand-side measures

Regulation and standards

- METI's Top Runner programme encourages the development of more energy-efficient products through continuous revisions of targets.
- Eco Action 21, an environmental management system for SMEs, was launched in 1996.
- MOE promotes environmental information disclosure through the Environmental Reporting Guideline and awards.
- Labelling to facilitate consumer choice including energy-saving labels and Eco-Mark scheme.

Public procurement and demand support

- The Law on Promoting Green Purchasing of 2000 requires all government institutions to implement green procurement.
- Support for the Green Purchasing Network (GPN) which facilitates green procurement by the private sector and citizen groups.

Sweden

Definition and strategy for eco-innovation

Definition of eco-innovation

- No specific definitions.

Strategy and initiatives for promoting eco-innovation

- The government formulated "Innovative Sweden" as a national innovation strategy in 2004 covering six sectors (automotive, IT/telecom, biotechnology, pharmaceuticals, metals, and pulp and paper).
- The government takes initiatives to assign governmental agencies to strengthen institutional structure for developing and incorporating environmental technologies and to inquire about strategic possibilities and factors.
- Swentec was established in 2008 to support governmental efforts in the area of environmental technologies.
- Nutek contributes to creating new enterprises and promoting sustainable economic growth and prosperity.

Environment in innovation policies

Overall priorities

- Climate change is one of the government's top environmental priorities.
- The Research and Innovation Bill 2009-10 provides a framework for central government-funded research and focuses on energy and climate change.
- With the latest budget bill, the focus on innovation policy shifts from grants to technology development and to measures for creating more efficient market.

Supply-side measures

Research and development

- VINNOVA supports R&D in the areas of engineering, transport, communications and working life to promote sustainable growth.
- MISTRA supports programmes that contribute to solving major environmental problems.
- FORMAS encourages and supports research related to sustainable development in the areas of the environment, agricultural sciences, fish and spatial planning.
- In the transport sector, the focus of R&D is on security and environmental issues.
- In the energy sector, Sweden participates in the Nordic Energy Research Programme and the new European Strategic Energy Technology (SET) Plan.
- The government co-finances the research project "Development of Cleaner Production" with the Swedish Environmental Research Institute (IVL).

Pre-commercialisation

- The Swedish Energy Agency (SEA) provides funds for pilot projects for the production of second-generation biofuels and several research programmes.
- Competence centres have been established for different technologies in the fields of renewable energy and energy efficiency.
- Nutek, VINNOVA and Innovationsbron organise incubation activities in environmental fields.
- The Research and Innovation Bill aims to promote the development and commercialisation of second-generation biofuels and new technologies for efficient vehicles and electricity production.
- The Delegation for Sustainable Cities provides subsidies to stimulate development of demonstration projects in the area of sustainable city building.

Education and training

- The Higher Education Law states that universities are responsible for promoting sustainable development.

Networking and partnership

- Sweden participates in European Technology Platforms, including the Hydrogen and Fuel Cell Technology platform and forest-based sector platform.

Information services

- The government will develop an EU catalogue of existing directories and databases on environmental technologies to disseminate case studies and results related to the use of environmental technologies.
- The Climate Investment Programme (Klimp) and the Local Investment Programme (LIP) were developed to raise public awareness of environmental issues.

Demand-side measures

Regulation and standards

- Swan Label, an official Nordic eco-label, has been in place since 1989 and covers over 60 groups of products.

Technology transfer

- SymbioCity was set up as a platform for Swedish companies exporting green technologies and sustainable construction to the world. Agreements on bilateral co-operation in the area of environmental technologies have been signed with China, Brazil, the United States, etc. In 2008, the government appointed a High Representative for Sino-Swedish Environmental Technology Co-operation.
- The government has tasked the Swedish Trade Council with promoting the export of environmental technologies, especially from SMEs.
- The government has tasked the Invest in Sweden Agency with promoting foreign investments in the environmental technology sector.

Co-ordination for eco-innovation

Policy co-ordination within government

- The Ministry of Enterprise, Energy and Communications works closely with the Ministry of the Environment, the Ministry of Foreign Affairs and the Ministry of Education and Research to boost knowledge of and skills for eco-innovation in the business sector.
- The public agencies VINNOVA, SEA, Nutek and Swentec work with government to facilitate eco-innovation.

Turkey

Definition and strategy for eco-innovation

Definition of eco-innovation

- Any form of innovation aiming at significant and demonstrable progress towards the goal of sustainable development, by reducing impacts on the environment or achieving a more efficient and responsible use of natural resources including energy.

Strategy and initiatives for promoting eco-innovation

- The 2006 National Rural Development Strategy aims to improve the management and development of protected areas.
- The National Environmental Strategy aims to support sustainable development and to meet people's need for a healthful environment.
- The Competitiveness and Innovation Framework Programme aims to encourage the competitiveness of SMEs by supporting innovation activities.

Environment in innovation policies

Overall priorities

- The Ministry of Environment and Forestry (MoEF) has the following priorities: reuse and recycling of wastewater, changes in consumption, integrated river basin management, determination of environmental quality standards and discharge standards for dangerous substances, and chemical and biological monitoring.

Supply-side measures

Research and development

- Notable environmental R&D projects include: integrated treatment of municipal wastewater and organic solid waste with renewable energy (bio-methane), recycling technologies, and tackling ozone-depleting substances.

Networking and partnership

- The Air Quality Monitoring Network was created to collect data on air emissions and quality, and benefits from efforts by provincial directorates and universities.

Information services

- The Technology Development Foundation of Turkey informs SMEs on phasing out the use of ozone-depleting substances in different sectors and on technology alternatives.
- MoEF's Biodiversity Monitoring Unit developed a biodiversity database, Prophet Noah's Ship.
- The Small and Medium Industry Development Organisation provides support mechanism for increasing the competitiveness of SMEs by encouraging entrepreneurship and innovative start-ups.

Demand-side measures

Regulation and standards

- The Environment Standards in the Textile Sector project is harmonising Turkish textile SMEs' practices with international environmental standards for materials testing.
- The Energy Efficiency Law of 2007 aims to increase efficiency awareness, training for energy managers and staff of future energy service companies and to improve administrative structures for energy efficiency services.
- The Pasture Law of 1998 aims for protection of biodiversity, sustainable use of pasture resources, and limiting land degradation and soil erosion.

Co-ordination for eco-innovation

Policy co-ordination within government

- The Ministry of Industry and Trade co-ordinates overall modalities of participation in EU projects such as the Competitiveness and Innovation Framework programme and the Entrepreneurship and Innovation programme.

United Kingdom

Definition and strategy for eco-innovation

Definition of eco-innovation

- The production, assimilation or exploitation of a novelty in products, production processes, services or in management and business methods, which aims to prevent or substantially reduce environmental risk, pollution and other negative impacts of resource use.
- Improvement in products and services come from innovations in business process, models, marketing as well as technologies.
- Any form of innovation contributes to sustainable development by reducing negative impacts on the environment, or achieving a more efficient and responsible use of resources.

Strategy and initiatives for promoting eco-innovation

- The Low Carbon Industrial Strategy will be developed in 2009 and set out the government's role in the development of low-carbon economy.
- The 2007 Commission on Environmental Markets and Economic Performance (CEMEP) brought together leaders from business, trade unions, universities and NGOs to develop recommendations on how the United Kingdom could exploit economic opportunities arising from the transition to a low-carbon, resource-efficient economy. The UK Low Carbon Industrial Strategy was released in July 2009.
- Supply-side initiatives include innovation platforms in the area of low-impact buildings and low-carbon vehicles, and an innovation white paper.

Environment in innovation policies

Overall priorities

- Eco-innovation relates to energy generation, sustainable consumption and production, low-carbon business opportunities, etc.

Supply-side measures

Equity support

- Tax incentives to support investment in innovative new technologies and high-risk ventures through R&D tax credit, the Enterprise Investment Scheme and the Venture Capital Trust.

Research and development

- The Technology Strategy Board (TSB) aims to stimulate innovation in the areas offering the greatest scope for boosting growth and productivity.
- The Energy Technologies Institute's technology programmes aim to accelerate the creation of innovative and commercially viable products and processes.

Pre-commercialisation

- The Environmental Transformation Fund focuses on the demonstration and deployment phases of bringing low-carbon and energy-efficient technologies to the market in the areas of low-impact buildings, assisted living and low-carbon vehicles.
- TSB's innovation platforms also aim to accelerate the development and commercialisation of early stage of radical technologies.
- Other technology demonstration programmes include those on hydrogen and fuel cell technologies, carbon abatement technologies, nanotechnology, and the Carbon Capture and Storage (CCS) Demonstration Competition.
- The Carbon Trust, a government company set up in 2001, works with organisations to develop commercial low-carbon technologies and businesses.

Education and training

- The Knowledge Transfer Partnership scheme funds graduates in science and engineering to work in innovative firms, including environmental firms.

Networking and partnership

- TSB organises innovation platforms in such areas as intelligent transport systems and services, low-impact buildings, assisted living, network security and low-carbon vehicles.
- The Centre of Excellence for Low Carbon and Fuel Cell Technologies aims to catalyse market transformation by linking technology providers and end users.
- The Energy Research Partnership brings people from energy research, development, demonstration and deployment in government, industry, academia and interested bodies together to identify and work towards shared goals.
- Knowledge Transfer Networks co-ordinated by TSB build capacity for innovation by promoting exchange of knowledge within and between sectors, helping SMEs access funding, and stimulating innovaion in their communities.

Information services

- The government funds the Energy Saving Trust which provides free information and advice and has a network of local advice centres throughout the country specifically designed to help companies and consumers take action to save energy.

Demand-side measures

Regulation and standards

- The Code for Sustainable Homes is helping to drive transformation in the housing market and catalyse innovation for zero-carbon homes.
- Incentives to encourage the adoption of new energy technologies include: stamp duty exemption for new zero-carbon homes; reduced VAT rate (5%) for the professional installation of micro-generation equipment in residential and charitable properties; exemption from climate change levy for supplies of electricity generated from renewable sources; exemption from income tax for surplus electricity sold by individual households; the Enhanced Capital Allowance scheme for energy and water efficient equipment.

Public procurement and demand support

- TSB takes an advisory role on public procurement to promote innovation in construction, food and business waste management. TSB's innovation platform also aims to leverage government procurement resources to increase business investment in R&D for innovation.
- Public procurement is referred to in the Low Carbon Transport Innovation Strategy and *Building a Green Future* policy statement.
- The Department for Business, Innovation and Skills (BIS) is supporting public procurers to apply the Forward Commitment Procurement model whereby procurers incentivise eco-innovation by agreeing to purchase at a specified future date and price an undefined product to solve a specified challenge with an environmental footprint smaller than current alternatives.
- All government departments must develop Innovative Procurement Plans by November 2009, including innovations for sustainability.

Technology transfer

- Sustainable Development Dialogues are encouraging transfer of industrial symbiosis techniques to Brazil, China, Mexico, etc.

Co-ordination for eco-innovation

Policy co-ordination within government

- The High-level Low Carbon Economy Policy Group, which was formed to follow up recommendations of CEMEP on eco-innovation, manages policy driving the transition to a more environmentally sustainable economy.

United States

Definition and strategy for eco-innovation

Definition of eco-innovation

- "Environmental innovation", "clean technology" (or "clean-tech") or "sustainable manufacturing" are the terms more often used.
- The Department of Commerce (DOC) defines sustainable manufacturing as the creation of manufactured products that use processes that are non-polluting, conserve energy and natural resources, and are economically sound and safe for employees, communities and consumers.

Strategy and initiatives for promoting eco-innovation

- DOC launched the Sustainable Manufacturing Initiative (SMI) and the Public-Private Dialogue to identify US industry's most pressing sustainable manufacturing challenges and to co-ordinate public and private sector efforts to address these challenges.
- The Environment Protection Agency (EPA) established the National Center for Environmental Innovation (NCEI) which promotes new ways to achieve better environmental results. It focuses on creating a results-oriented regulatory system, promoting environmental stewardship across society, and building capacity for innovative problem-solving.
- EPA laid out *Innovating for Better Environmental Results: A Strategy to Guide the Next Generation of Innovation* in 2002. This strategy is directed at the agency's own policy innovation.

Environment in innovation policies

Overall priorities

- The government addresses the innovation perspective especially in the field of climate change, air pollution and energy.
- Foster multiple forms of collaboration within and across agencies, with industry, academic, non-profit organisations and states.

- Clearer orientation towards problem solving and focus on dissemination and commercialisation of environmental technologies.

Supply-side measures

Equity support

- The Small Business Innovation Research programme provides grant funding to small businesses for developing innovative technologies with a focus on proof of concept and commercial prototype.
- The Technology Commercialization Fund (TCF) targets supporting early-stage product development and makes matching funds available to private sector partners.

Research and development

- The Department of Energy (DOE)'s Hydrogen, Fuel Cells and Infrastructure Technologies Program focuses on the development of next-generation technologies, establishment of an education campaign that communicates potential benefits, and better integration of sub-programmes in hydrogen, fuel cells and distributed energy.
- All technologies developed by DOE must meet environmental regulations.

Pre-commercialisation

- The EPA's R&D Continuum describes the progression of technology development from idea through diffusion in the market.
- The DOE's Technology Innovation Program supports commercialisation of emerging technologies.

Education and training

- The Green Engineering Program aims to incorporate risk-related concepts into chemical processes and products designed by academia and industry. It developed a textbook for engineering educators and continuing education courses for engineers.

Networking and partnership

- EPA's Design for the Environment Program works in partnership with a broad range of stakeholders to reduce risk to people and the environment by preventing pollution.

- DOE's Lawrence Livermore National Laboratory performs key research in water and environment, energy, carbon and climate in collaboration with 80 universities, companies and research organisations.

Information services

- The EPA created the Environmental Technology Opportunities Portal to match companies and organisations with programmes for fostering environmental technologies and to relay information on EPA's technologies for air, water, and waste treatment and control.
- The DOC's SMI and Public-Private Dialogue established a web portal for companies that provide information on what DOC and other federal agencies are doing to support sustainable manufacturing.

Demand-side measures

Public procurement and demand support

- Since 1993, the government has aimed to strengthen federal agencies' environmental, energy and transport management. This includes the requirement for federal agencies to apply sustainable practices when acquiring goods and services, including the purchasing of bio-based, environmentally preferable, energy-efficient, water-efficient and recycled-content products.
- EPA and the General Services Administration help agencies find environmentally preferable products by providing online guidance and a catalogue.
- The Energy Independence and Security Act promotes the purchase of energy-efficient products and alternative fuels by federal agencies. The Federal Electronics Challenge promotes agencies' purchase of electronics that meet certain environmental criteria.

Technology transfer

- EPA supports the promotion of exports in clean, efficient energy technologies to India, China and other developing countries.
- The Clean Energy Technology Export Program is a public-private partnership for addressing export barriers in the world clean technology market.

- The Environmental Exports Program helps mitigate risk for US companies and offers competitive financing terms to international buyers for the purchase of US environmental goods and services.

Co-ordination for eco-innovation

Policy co-ordination within government

- DOC's Manufacturing and Services Unit created an interagency working group on sustainable manufacturing under the Interagency Working Group on Manufacturing Competitiveness, which brings together more than 17 agencies.

Chapter 6

Looking Ahead: Key Findings and Prospects for Future Work on Sustainable Manufacturing and Eco-Innovation

This chapter draws together the findings from the previous five chapters into nine key messages. It identifies promising areas for the next phases of the OECD project on sustainable manufacturing and eco-innovation and presents the recommendations from the project's advisory expert group. These include two major areas of work: i) *improving the clarity and consistency of sustainable manufacturing indicators to support industry efforts; and* ii) *filling gaps in the understanding of eco-innovation through case studies and guiding innovative policy making by sharing best practices and long-term visions as well as benchmarking.*

Introduction

The preceding chapters have presented the results of research and analysis carried out during the first phase of the OECD project on sustainable manufacturing and eco-innovation. The aim of the project is to help accelerate sustainable production efforts by manufacturing industries and to promote the concept of eco-innovation in order to invigorate new technological and systemic solutions to global environmental challenges. The project initially focused on helping policy makers and industry practitioners understand relevant concepts and practices and on setting directions for future work to fill gaps in understanding and analysis. For this purpose, the following research activities were undertaken:

- review the concepts of sustainable manufacturing and eco-innovation and build a framework for analysis;
- analyse eco-innovation processes on the basis of existing examples from manufacturing companies.
- benchmark the sets of indicators used by industry to achieve sustainable manufacturing.
- analyse the strengths and weaknesses of existing methodologies for measuring eco-innovation at the macro level.
- take stock of national strategies and policy initiatives to promote eco-innovation in OECD countries.

This concluding chapter draws together the findings from the preceding chapters into nine key points. Based on the research outcomes, promising areas of work for the project's next phases are presented, as identified by the project's advisory expert group.

Nine key findings

1. Practices for sustainable manufacturing have evolved

Manufacturing industries have the potential to become a driving force for realising a sustainable society by introducing efficient production practices and developing products and services that contribute to better environmental performance. Driven in part by stricter environmental regulations, manufacturing industries have applied various control and treatment measures to reduce the amount of emissions and effluents. In recent years, their efforts to improve environmental performance have shifted from such end-of-pipe solutions to a focus on product life cycles and integrated environmental strategies and management systems, as many companies are beginning to accept larger environmental and social responsibilities throughout their value chain.

Furthermore, efforts are increasingly made to create closed-loop, circular production systems which regenerate discarded products as new resources for production. For example, the establishment of eco-industrial parks aims at harnessing economic and environmental synergies between traditionally unrelated manufacturers. The adoption of more integrated and systematic methods to improve sustainability performance has also laid the foundation for new business models or modes of provision that do not need to rely on intensive use of natural resources to make profits.

2. Eco-innovation seeks more radical improvements

Much attention has recently been paid to innovation as a way for industry and policy makers to work towards more radical improvements in corporate environmental practices and performance. Many companies have started to use *eco-innovation* or similar terms to describe their contributions to sustainable development. A few governments are also promoting the concept as a way to meet sustainable development targets while keeping industry and the economy competitive.

The European Union (EU) considers eco-innovation as a way to support the wider objectives of its Lisbon Strategy for competitiveness and economic growth. The concept is promoted primarily through the Environmental Technology Action Plan (ETAP), which defines eco-innovation as "the production, assimilation or exploitation of a novelty in products, production processes, services or in management and business methods, which aims, throughout its life cycle, to prevent or substantially reduce environmental risk, pollution and other negative impacts of resource use (including energy)". Environmental technologies are also considered to have promise for improving environmental conditions without impeding economic growth in the United States, where they are promoted through various public-private partnership programmes and tax credits (OECD, 2008).

To date, the promotion of eco-innovation has focused mainly on environmental technologies, but there is a trend to broaden the scope of the concept. In Japan, the government's Industrial Science Technology Policy Committee defines eco-innovation as "a new field of techno-social innovations [that] focuses less on products' functions and more on [the] environment and people" (METI, 2007). Eco-innovation is here seen as an overarching concept which provides direction and vision for pursuing the overall societal changes needed to achieve sustainable development. This extension of eco-innovation's scope corresponds to the more integrated application of sustainable manufacturing described above.

3. Eco-innovation has three dimensions: targets, mechanisms and impacts

The definition of innovation in the OECD's *Oslo Manual*[1] generally applies to eco-innovation, but eco-innovation has two further significant, distinguishing characteristics:

- Eco-innovation represents innovation that results in a reduction of environmental impact, whether that effect is intended or not.
- The scope of eco-innovation may go beyond the conventional organisational boundaries of the innovating organisation and involve broader social arrangements that trigger changes in existing socio-cultural norms and institutional structures.

These features lead to a new understanding of eco-innovation in terms of:

- **targets**, which are the basic focus of eco-innovation. These are *products* (goods and services), *processes*, *marketing methods*, *organisations*, and *institutions* (institutional arrangements and socio-cultural norms). Eco-innovation in products and processes tends to rely on *technological* development, while eco-innovation in marketing, organisations and institutions relies more on *non-technological* changes.
- **mechanisms**, which are how changes in the target areas are made. They can involve *modification* and *redesign* of practices, *alternatives* to existing practices, or the *creation* of new practices. It is also associated with the underlying nature of the eco-innovation – whether the change is of a technological or non-technological character.
- **impacts**, which are how the eco-innovation affects environmental conditions across product life cycles or other dimensions. Experience shows that more radical changes, such as alternatives and creation, usually have the potential for higher environmental benefits.

4. Sustainable manufacturing calls for multi-level eco-innovations

Innovation plays a key role in moving manufacturing industries towards sustainable production. Evolving sustainable manufacturing initiatives – from traditional pollution control through cleaner production initiatives, to life cycle thinking and the establishment of closed-loop production – can be viewed as facilitated by eco-innovation. While more integrated approaches such as closed-loop production can potentially yield higher environmental improvements, they need to involve a combination of a wider range of innovation targets and mechanisms to leverage their benefits. As sustainable manufacturing initiatives advance, the nature of the eco-innovation process becomes increasingly complex and more difficult to co-ordinate.

These advanced, multi-level eco-innovation processes are often referred to as *system innovation* – an innovation characterised by fundamental shifts in how society functions and how its needs are met (Geels, 2005). System innovation may have its source in technological advances, but technology alone will not make a great difference. It has to be associated with organisational and social structures and with human and cultural values. While this may indicate the difficulty of achieving large-scale environmental improvements, it also hints at the need for manufacturing industries to adopt an approach that aims to integrate the various elements of the eco-innovation process so as to leverage the maximum environmental benefits.

5. Current eco-innovations focus mostly on technological development but are facilitated by non-technological changes

According to a review of eco-innovation examples from three sectors (automotive and transport, iron and steel, and electronics), the primary focus of current eco-innovation in manufacturing industries tends to rely on technological advances. These are typically associated with product or process as the eco-innovation targets, and with modification or redesign as the principal mechanisms. Nevertheless, even with a strong focus on technology, a number of complementary non-technological changes have functioned as key drivers for these developments. Such changes have been either organisational or institutional in nature, including the establishment of separate environmental divisions or multi-stakeholder collaborative research networks. Some industry players have also started exploring more systemic eco-innovation through new business models and alternative modes of provision such as a bicycle-sharing scheme in Paris and product-service solutions in photocopying and data centre energy management.

Hence, the heart of an eco-innovation cannot necessarily be represented adequately by a single set of target and mechanism characteristics. Instead, eco-innovation seems best examined and developed using an array of characteristics ranging from modification to creation across products, processes, organisations and institutions.

6. Clear and consistent indicators are needed to accelerate corporate sustainability efforts

Indicators help manufacturing companies to understand environmental issues surrounding existing production systems, define specific objectives and monitor progress towards sustainable production. There are many available indicators for sustainable manufacturing. They are diverse in nature, and have been developed on a voluntary basis, or as a standard or as part of legislation.

Among the nine representative sets of indicators reviewed (individual indicators, key performance indicators, composite indices, material flow analysis, environmental accounting, eco-efficiency indicators, life cycle assessment, sustainability reporting indicators, and socially responsible investment indices), there is no one set of indicators that covers all aspects that manufacturing companies need to consider for sustainable production. Many are applying more than one set of indicators for management decision making and operational improvement, often without relating them.

An appropriate combination of existing indicator sets could help give companies a more comprehensive picture of economic, environmental and social effects across the value chain and product life cycle. The further development and standardisation of environmental valuation techniques could also help companies make more rational decisions on investments in sustainable manufacturing activities. Life cycle considerations have helped companies to consider environmental effects beyond their factory gates, but new system-level indicators may also need to be developed to identify the wider impacts of introducing new products and production processes beyond a single product life cycle.

7. Improved benchmarking and better indicators would help deepen understanding of eco-innovation

Quantitative measurement of eco-innovation activities would improve understanding of the concept and practices and help policy makers analyse trends. It would also raise awareness of eco-innovation among industry, policy makers and other stakeholders, and would make improvements achieved through eco-innovation more evident to producers and consumers alike.

To improve understanding of the diversity and characteristics of eco-innovation activities for better policy making, the nature (how companies innovate), drivers and barriers, and impacts of eco-innovation need to be captured at the macro (sectoral, local and national) level. Those aspects can be measured and analysed by using four categories of data: input measures (*e.g.* R&D expenditure); intermediate output measures (*e.g.* number of patents); direct output measures (*e.g.* number of new products); and indirect impact measures (*e.g.* changes in resource productivity). Relevant data can be obtained either by using generic data sources or by conducting specially designed surveys.

Each measurement approach has its strengths and weaknesses, and no single method or indicator can capture eco-innovation comprehensively. Generic data sources can provide readily available information on certain aspects of the nature of eco-innovation, but the scope of analysis may be limited. While surveys could enable researchers to obtain more detailed and focused information on different aspects of eco-innovation, they are costly to conduct and the number of respondents would be limited.

To identify overall patterns of eco-innovation, it is important to apply different analytical methods, possibly combined, and view various indicators together. The development of an "eco-innovation scoreboard", which combines statistics from generic data sources, could greatly improve government and industry awareness of eco-innovation by benchmarking the progress of national efforts. Measuring the "greenness of national innovation systems" could offer another avenue for benchmarking eco-innovation and could be linked to a scoreboard.

8. To promote eco-innovation, integration of innovation and environmental policies is crucial

Stringent environmental regulations and standards do not give firms enough incentive to innovate beyond end-of-pipe solutions, although they have helped to reduce environmental damage to a large extent. Recently, market-oriented instruments, such as green taxes and tradable permits, have been introduced as more efficient measures to trigger the development and deployment of green technologies. Yet, to realise its potential, eco-innovation will require actions to ensure that the full cycle of innovation is efficient, with policies ranging from investments in R&D to support for demonstrating and commercialising existing and breakthrough technologies.

Innovation policy, on the other hand, has focused on spurring economic growth by developing new technologies for improving productivity and new areas of functionality. It has been too broad to address specific environmental concerns appropriately. Eco-innovation has not been a primary objective of environmental or of innovation policy.

Both policy areas would benefit from closer integration. More innovation-oriented environment policy could make improvements in environmental quality more attainable through better application of technologies, while reducing the costs of environmental measures and benefiting from new market opportunities in a growing eco-industry. From the innovation point of view, it is increasingly recognised that "third-generation innovation policies have to become fully horizontal and support a broad range of social goals if they are to achieve their objective of increasing the overall innovation rate in societies" (OECD, 2005, p. 57).

9. Creating successful eco-innovation policy mixes requires understanding the interaction of demand and supply

Results of a survey of ten OECD governments (Canada, Denmark, France, Germany, Greece, Japan, Sweden, Turkey, the United Kingdom and the United States) on current national strategies and policy initiatives for eco-innovation show that an increasing number of countries now perceive environmental challenges not as a barrier to economic growth but as a new opportunity for increasing competitiveness. But not all countries surveyed seem to have a specific strategy for eco-innovation; when they do, there is often little policy co-ordination among the various departments involved.

Government policy initiatives and programmes that promote eco-innovation are diverse and include both supply-side and demand-side measures. As most countries recognise the need for a more collaborative approach to developing the technologies required to face today's environmental challenges, many supply-side initiatives involve creating networks, platforms or partnerships that engage different industry and non-industry stakeholders, in addition to conventional measures for funding research, education and technology demonstration.

Demand-side measures are receiving increasing attention, as governments acknowledge that insufficiently developed markets are often the key constraint for eco-innovation. For example, green public procurement provides an opportunity to foster demand for eco-innovation, although such policies need to be carefully designed not to harm competition or support technologies with sub-optimal performance. Current demand-side measures are often poorly aligned with existing supply-side measures and require a more focused approach to leveraging eco-innovation activities. A more comprehensive understanding of the interaction between supply and demand for eco-innovation will be a pre-requisite for creating successful eco-innovation policy mixes.

Main lessons of the first phase

In order to meet global environmental challenges such as climate change, increasing attention has been paid to innovation as a way to develop sustainable solutions. The concepts of sustainable manufacturing and eco-innovation are increasingly adopted by industry and policy makers as a way to facilitate more radical and system-wide improvement in production processes and products/services and in corporate environmental performance.

To date, the primary focus of sustainable manufacturing and eco-innovation tends to be on technological advances for the modification and redesign of products or processes, as in the case of conventional innovation. However, some advanced industry players have adopted complementary organisational or institutional changes such as new business models or alternative modes of provision, for example, offering product-service solutions rather than selling only physical products.

An appropriate combination of existing sets of indicators could help businesses gain a more comprehensive picture of environmental effects across the value chain and product life cycle. Clearer and more consistent indicators would increase their ability to manage and improve environmental performance. Indicators should also be made applicable for supply chain companies and small and medium-sized enterprises (SMEs) in order to facilitate life cycle-wide improvements.

Quantitative measurement can also help policy makers and industry better grasp overall trends and the characteristics of eco-innovation. Since no single measurement approach can capture eco-innovation comprehensively, it is important to apply different analytical methods, possibly in combination, and view different indicators together.

Closer integration of innovation and environmental policies could benefit both policy areas and accelerate corporate efforts on sustainable manufacturing and eco-innovation. Survey results show that not all countries have a specific eco-innovation strategy. For those that do, there is limited policy co-ordination between different government departments. Current policy initiatives and programmes are diverse and include both supply-side and demand-side measures. A more comprehensive understanding of the interaction of supply and demand for eco-innovation is necessary to create a successful eco-innovation policy mix.

Prospects for future work

Based on the above research outcomes, promising work areas for the OECD project on sustainable manufacturing and eco-innovation in the next phase were identified as follows:

- **Provide guidance on indicators for sustainable manufacturing**. The OECD could bring clarity and consistency to existing indicator sets by working with other stakeholders on developing a common terminology and understanding of the indicators and their usage. It could also play a role in providing supportive measures for increasing the use of indicators by supply chain companies and SMEs. Further down the line, the OECD could utilise its experience in leading the development of the Pollutant Release and Transfer Register (PRTR) system[2] to standardise indicator sets and the methodology for both the micro level (facility, product or company) and the macro level (sectoral, local or national). To encourage system innovations, a framework for identifying system-wide impacts of new products and production processes could also be considered.
- **Identify promising eco-innovation policies**. Better evaluation of the implementation of different policy measures for eco-innovation would help to identify promising eco-innovation policies as well as the contexts in which specific policy instruments can be deployed effectively. The OECD can facilitate the sharing of best policy practices in this area among governments.
- **Build a common vision for eco-innovation**. The OECD could help fill the gap in understanding eco-innovations, especially those that are more integrated and systemic and have non-technological characteristics, by co-ordinating in-depth case studies. To guide industry and policy makers towards more radical and system-wide improvements, it could also work on the development of a common vision of environmentally friendly social systems and roadmaps for realising them. This exercise should involve member countries, industry experts, academics and NGOs.
- **Develop a common definition and a scoreboard**. Building upon its experience with innovation measurement and the *Oslo Manual*, the OECD could consider developing a common definition of eco-innovation and an "eco-innovation scoreboard" for benchmarking eco-innovation activities and public policies by combining different statistics and data. Such work could improve awareness and guide government efforts.

The project's advisory expert group also recommended conducting the following activities for the next phase of work:

Sustainable manufacturing indicators

- Develop a toolbox or manual to help manufacturing companies use existing indicator sets to improve their environmental performance by providing guidance and general recommendations on terminology, standard processes, methodologies and the use of indicators.
- Standardise methodologies for material flow analysis at the micro level (*i.e.* at the facility, corporate and product level), as this is considered one of the most effective tools for improving energy and resource efficiency.

Global eco-innovation platform

- Collect interesting examples of different levels of eco-innovation from around the world and conduct an in-depth study on processes that help achieve eco-innovation in order to draw lessons for practitioners and policy makers.
- Collect good examples of policies that promote eco-innovation and conduct an in-depth study on how they function. This can be followed by the identification of results-oriented, dynamic new-generation innovation policies that encourage industry to lead eco-innovation efforts.
- The above industry and policy best practices could be compiled as a freely accessible online database for knowledge sharing and networking as well as shared through workshops, conferences, etc.

Notes

1. Innovation is defined as "the implementation of a new or significantly improved product (good or service), or process, a new marketing method, or a new organisational method in business practices, workplace organisation or external relations" (OECD and Eurostat, 2005, p. 46).

2. PRTR is a national or regional environmental database or inventory of hazardous chemical substances and pollutants released to air, water and soil, and transferred off-site for treatment or disposal. Individual facilities determine, collect and report their releases and transfers to a national PRTR. Industry can benefit from PRTR data, as they can verify their own data by comparing it with others engaged in the same business activity. PRTR reporting may also contribute to industry identifying leaks, reducing waste and thereby saving money. The OECD began work on PRTRs in response to Agenda 21 adopted at the United Nations Conference on Environment and Development (UNCED) in Rio de Janeiro in 1992. In 1996, the OECD Council adopted a Recommendation on Implementing Pollutant Release and Transfer Registers [C(96)41/FINAL, amended as C(2003)87 in 2003], which called for its member countries to establish a PRTR. By 2007, 17 OECD countries established an operational PRTR and more are in a process of developing such a system (OECD, 2007).

References

Geels, F.W. (2005), *Technological Transitions and System Innovations: A Co-evolutionary and Socio-technical Analysis*, Edward Elgar, Cheltenham.

Ministry of Economy, Trade and Industry, Japan (METI) (2007), *The Key to Innovation Creation and the Promotion of Eco-Innovation*, report by the Industrial Science Technology Policy Committee of the Industrial Structure Council, METI, Tokyo.

OECD (2005), *Governance of Innovation Systems, Volume 1: Synthesis Report*, OECD, Paris.

OECD (2007), Pollutant Release and Transfer Register (PRTR) brochure, OECD, Paris, *www.oecd.org/dataoecd/35/26/39785042.pdf.*

OECD (2008), "Environmental Innovation and Global Markets", report for the Working Party on Global and Structural Policies, Environment Policy Committee, OECD, Paris, *www.olis.oecd.org/olis/2007doc.nsf/linkto/env-epoc-gsp(2007)2-final.*

OECD and Statistical Office of the European Communities (Eurostat) (2005), *Oslo Manual: Guidelines for Collecting and Interpreting Innovation Data*, 3rd edition, OECD, Paris.

Glossary

Cleaner production	A preventive approach to the production process which aims to minimise the input of energy and materials and the quantity and toxicity of emissions and wastes at the source rather than at the end of the process.
Closed-loop production	A method of production which aims to achieve a closed material resource cycle in which all components in the production system are reused, remanufactured or recycled as new input.
Corporate social responsibility (CSR)	The idea that companies should take social and environmental concerns as well as their economic goals and regulatory responsibilities into account in their operations.
Eco-efficiency	A concept promoting the efficient use of resources and less generation of waste and pollution in economic activities. Eco-efficiency can be measured as economic value created per unit of environmental impact (or *vice versa*).
Eco-industrial park	A cluster of companies that co-operate closely with each other and with the local community to share resources to improve economic performance and minimise waste and pollution. The collective benefit is considered greater than the sum of the benefits companies would realise when optimising only their individual performance.
Eco-innovation	Innovation which, intentionally or not, results in a reduction of environmental impact compared to relevant alternatives.
End-of-pipe technology	Technology used to reduce already formed contaminants prior to discharge into the environment, as opposed to technology to reduce resource use and prevent pollution in the first place.

Environmental management system (EMS)	A way for organisations to implement their environmental management effectively and continuously improve their environmental performance, based on pre-determined objectives and targets. It normally consists of a cycle consisting of four steps: planning, implementing, monitoring/checking and reviewing/improving.
Green public procurement	A practice whereby public agencies include environmental criteria in tendering procedures for goods, services or works as a way to use their purchasing power to nurture a market for environmentally sound products.
Green tax	A tax intended to make the choices and activities of producers and/or consumers more environmentally sound by internalising some of the cost of environmental impacts which are not conventionally accounted for in the market price.
Industrial ecology	A framework to design and operate industrial activities in harmony with ecological systems through extensive application of closed-loop production beyond the boundary of a single company.
Innovation	The implementation of a new or significantly improved product (good or service), or process, a new marketing method, or a new organisational method in business practices, workplace organisation or external relations.
Innovation system	A concept which stresses the flow of technology and information among people, enterprises and institutions as the means of turning an idea into an innovation that is successfully deployed in the market.
Institutional innovation	Innovation characterised by institutional changes, including changes in the roles and structures of industry and public institutions, infrastructures, relationships with other organisations, laws and rules, social norms and practices, cultural values, patterns of behaviour, etc.
Kyoto Protocol	An international agreement (adopted in 1997) that sets binding targets for industrialised countries to reduce greenhouse gas emissions by an average of 5% against 1990 levels over the five-year period 2008-12.

Life cycle assessment (LCA) A method for assessing the overall environmental impact of a product, a process or a service over its entire life cycle, *i.e.* from extraction of resources through processing and production to use and disposal. This normally involves the collection and evaluation of quantitative data on the inputs and outputs of materials, energy and waste flows.

Manufacturing industries Industry sectors which transform materials or components into new products which are either sold to customers or components used in other production processes.

Mass balance (material balance) An analytical concept which helps to understand the flow of materials through a system (process, facility, industry or geographical region). Because of the fundamental physical principle that matter is neither created nor destroyed, the material input from the environment into a system balances the output from the system as products, emissions and wastes, plus any change in stocks. By examining the difference between input and output, material flows which might have been unknown or difficult to measure can be identified.

Non-technological innovation Innovation characterised by changes in the structures or functioning of organisations/institutions, management practices, marketing methods, business models, etc.

Product-service system (PSS) A business model that focuses on delivering consumer utility rather than the production and supply of physical goods. The use of a service to meet certain consumer needs is considered a way to lower the environmental impact of the products involved.

Remanufacturing An industrial process in which used products are restored to a condition equivalent to the original products. Normally, used products are disassembled and useable parts are cleaned or refurbished. New products are manufactured by reassembling refurbished parts with new parts where necessary.

Research and development (R&D) Creative work undertaken on a systematic basis in order to increase the stock of knowledge, including knowledge of man, culture and society, and the use of this stock of knowledge to devise new applications. It covers three areas of activities: basic research, applied research and experimental development.

Sustainability reporting
A practice by which organisations measure and disclose their impact on and contribution to economic, environmental and social conditions. It can help them manage their efforts to reach the goal of sustainable development and also improve transparency and accountability to stakeholders.

Sustainable consumption
The use of products and services that meet basic needs, improve quality of life, minimise the use of natural resources and toxic materials, and reduce emissions of waste and pollutants so as not to jeopardise the needs of future generations.

Sustainable production
The creation of goods and services using processes and systems that reduce the use of natural resources and toxic materials and emissions of waste and pollutants, protect workers, communities and consumers, and are economically viable.

System innovation
Innovation to achieve major changes in how societal functions and needs such as transport, communication, housing, feeding, and energy are fulfilled. It typically involves the concomitant evolution of technological solutions, infrastructures, social practices, regulations and industry structures.

Tradable permit
Right to sell and buy actual or potential pollution in artificially created markets. This is used as a market-driven scheme to reduce emissions such as greenhouse gas (GHG) and sulphur dioxide (SO_2). Authorities set the limit for the emission of a particular gas and allocate the emissions quota to individual companies. If companies emit less than their quota they can sell their permits; if they emit more than their quota they have to buy permits from other companies. This "cap and trade" scheme is considered to encourage companies to pollute as little as possible.

References

Online references

Center for Environmental Economic Development (CEED) Innovative Approaches to Revenue website, Arcata, CA, *www.ceedweb.org*.

Economy Watch website, Singapore, *www.economywatch.com*.

European Commission (EC) Corporate Social Responsibility website, DG Enterprise and Industry, Brussels, *http://ec.europa.eu/enterprise/csr/index_en.htm*.

EC Green Public Procurement glossary, DG Environment, Brussels, *http://ec.europa.eu/environment/gpp/glossary_en.htm*.

Lowell Center for Sustainable Production (LCSP) website, University of Massachusetts Lowell, Lowell, MA, *http://sustainableproduction.org*.

Mass Balance UK project website, Biffaward, Newark, Nottinghamshire, *www.massbalance.org*.

OECD Glossary of Statistical Terms, *http://stats.oecd.org/glossary/index.htm*.

United Nations Environment Programme (UNEP) Cleaner Production website, UNEP, Paris, *www.unep.fr/scp/cp*.

United Nations Framework Convention on Climate Change (UNFCCC) website, Bonn, *www.unfccc.int*.

United Nations Statistics Division Environment Glossary website, UNSD, New York, *http://unstats.un.org/unsd/ENVIRONMENTGL/default.asp*.

Other references

Geels, F.W. (2004), "Understanding System Innovations: A critical literature review and a conceptual synthesis", in B. Elzen, F.W. Geels and K. Green (eds.), *System Innovation and the Transition to Sustainability: Theory, evidence and policy*, pp. 19-47, Edward Elgar, Cheltenham.

Global Reporting Initiative (GRI) (2006), *GRI Sustainability Reporting Guidelines* (3rd ed.), GRI, Amsterdam, *www.globalreporting.org*.

Levine, R.S., M.T. Hughes and C.R. Mather (n.d.), *Thesaurus of Sustainability*, Center for Sustainable Cities, University of Kentucky, Lexington, KY, *www.cscdesignstudio.com*.

Lowe, E.A. (2001), *Eco-industrial Park Handbook for Asian Developing Countries*, A report to the Asian Development Bank, Indigo Development, Oakland, CA, *www.indigodev.com/Ecoparks.html.*

Lund, R. (1998), *Remanufacturing: An American Resource*, Proceedings of the Fifth International Congress for Environmentally Conscious Design and Manufacturing, 16-17 June, Rochester Institute of Technology, Rochester, NY.

Norwegian Ministry of the Environment (1994), Oslo Roundtable on Sustainable Production and Consumption, IISD, Winnipeg, *www.iisd.ca/consume/oslo000.html.*

OECD (1997), *National Innovation Systems*, OECD, Paris.

OECD (2002), *Frascati Manual: Proposed Standard Practice for Surveys on Research and Experimental Development*, OECD, Paris.

OECD and Statistical Office of the European Communities (Eurostat) (2005), *Oslo Manual: Guidelines for Collecting and Interpreting Innovation Data*, OECD, Paris.

UNEP (2002), *Product-Service Systems and Sustainability: Opportunities for Sustainable Solutions*, UNEP, Paris, *www.unep.fr/scp/design/pss.htm.*

OECD PUBLISHING, 2, rue André-Pascal, 75775 PARIS CEDEX 16
PRINTED IN FRANCE
(92 2009 06 1 P) ISBN 978-92-64-07721-8 – No. 57117 2009